Chitosan Nanoparticles

Chitosan Nanoparticles

Edited by:

Abhishek Gupta

ARCLER PRESS

www.arclerpress.com

Chitosan Nanoparticles

Abhishek Guptan

Arcler Press
2010 Winston Park Drive,
2nd Floor
Oakville, ON L6H 5R7
Canada
www.arclerpress.com
Tel: 001-289-291-7705
001-905-616-2116
Fax: 001-289-291-7601
Email: orders@arclereducation.com

ISBN: 978-1-77361-209-6 (Hardcover)

DECLARATION

Some content or chapters in this book are open access copyright free published research work, which is published under Creative Commons License and are indicated with the citation. We are thankful to the publishers and authors of the content and chapters as without them this book wouldn't have been possible.

ABOUT THE EDITOR

Dr. Abhishek Gupta is currently a National Post-Doctoral Fellow (NPDF) at All India Institute of Medical Sciences (AIIMS), New Delhi. He has completed their Doctoral Degree (Ph.D) from School of Basic Sciences, Indian Institute of Technology (IIT) Mandi HP India. He obtained his B.Sc. in Biological Sciences from MJP R University Bareilly India and M.Sc. Biotechnology from School of Biotechnology, Devi Ahilya University, Indore India. His recent research is focused on carbon nanoparticles (C-dots) for toxic metal and biomolecule sensing system. He has more than 15 publications in reputed international journal (Nano Letter, Chem Comm, JPC Letter, ACS Applied Materials and Interfaces, Nature Scientific Reports and etc.) with average Impact factor > 5.50

TABLE OF CONTENTS

Section II

LIST OF CONTRIBUTORS

Abhishek Gupta

Arezou Ghadi
Faculty of Chemical Engineering, Noshirvani University of Technology, Babol, Iran

Soleiman Mahjoub
Fatemeh Zahra Infertility and Reproductive Health Research Center, Babol University of Medical Sciences, Babol, Iran
Department of Biochemistry & Biophysics, Faculty of Medicine, Babol University of Medical Sciences, Babol, Iran

Fatemeh Tabandeh
National Institute of Genetic Engineering and Biotechnology (NIGEB), Tehran, Iran

Farid Talebnia
Faculty of Chemical Engineering, Noshirvani University of Technology, Babol, Iran.

Vivek Kamat
Nanobioscience, Agharkar Research Institute, GG Agarkar Road, Pune 411 004 India

Dhananjay Bodas
Nanobioscience, Agharkar Research Institute, GG Agarkar Road, Pune 411 004 India

Kishore Paknikar
Nanobioscience, Agharkar Research Institute, GG Agarkar Road, Pune 411 004 India

Reynaldo Esquivel
Departamento de Investigación en Física, Universidad de Sonora, Rosales y Transversal, 83000 Hermosillo, SON, Mexico

Josué Juárez
Departamento de Investigación en Física, Universidad de Sonora, Rosales y Transversal, 83000 Hermosillo, SON, Mexico

Mario Almada
Departamento de Investigación en Física, Universidad de Sonora, Rosales y Transversal, 83000 Hermosillo, SON, Mexico

Jaime Ibarra
Departamento de Investigación en Física, Universidad de Sonora, Rosales y Transversal, 83000 Hermosillo, SON, Mexico

Miguel A. Valdez
Departamento de Investigación en Física, Universidad de Sonora, Rosales y Transversal, 83000 Hermosillo, SON, Mexico

Tan Vui Nguyen
Institute of Biotechnology and EnvironmentTay Nguyen UniversityBuon Ma ThuotVietnam

Thi Thu Ha Nguyen
Institute of Biotechnology and EnvironmentTay Nguyen UniversityBuon Ma ThuotVietnam

San-Lang Wang
Department of Chemistry, Life Science Development CentreTamkang UniversityNew TaipeiTaiwan

Thi Phuong Khanh Vo
Institute of Biotechnology and EnvironmentTay Nguyen UniversityBuon Ma ThuotVietnam

Anh Dzung Nguyen
Institute of Biotechnology and EnvironmentTay Nguyen UniversityBuon Ma ThuotVietnam

Sijumon Kunjachan
Department of Pharmaceutical Sciences, Mahatma Gandhi University, Kerala, India
Department of Experimental Molecular Imaging (EXMI), University Clinic, RWTH Aachen, Germany

Sajan Jose
Department of Pharmaceutical Sciences, Mahatma Gandhi University, Kerala, India

Twan Lammers
Department of Experimental Molecular Imaging (EXMI), University Clinic, RWTH Aachen, Germany

M.AbdElgadir
Department of Pharmacology and Chemistry, Faculty of Pharmacy, Universiti Teknologi MARA, 42300 Bandar Puncak Alam, Selangor, Malaysia

Md.SalimUddin
Department of Pharmaceutical Technology, Faculty of Pharmacy, International Islamic University Malaysia, Kuantan Campus, 25200 Kuantan, Pahang, Malaysia

SahenaFerdosh
Faculty of Science, International Islamic University Malaysia, Kuantan Campus, 25200 Kuantan, Pahang, Malaysia

AishahAdam
Department of Pharmacology and Chemistry, Faculty of Pharmacy, Universiti Teknologi MARA, 42300 Bandar Puncak Alam, Selangor, Malaysia

Ahmed Jalal KhanChowdhury
Faculty of Science, International Islamic University Malaysia, Kuantan Campus, 25200 Kuantan, Pahang, Malaysia

Md.Zaidul IslamSarker
Department of Pharmaceutical Technology, Faculty of Pharmacy, International Islamic University Malaysia, Kuantan Campus, 25200 Kuantan, Pahang, Malaysia

Débora P Facchi
Environmental Engineering Program, Federal University of Technology, Paraná, Brazil

Suelen P Facchi
Environmental Engineering Program, Federal University of Technology, Paraná, Brazil

Alessandro F Martins
Environmental Engineering Program, Federal University of Technology, Paraná, Brazil
Materials Science & Engineering Program, Federal University of Technology, Paraná, Brazil

Patel R
Ramanbhai Patel College of Pharmacy, Charotar University of Science and Technology, CHARUSAT Campus, Changa – 388421, Anand, Gujarat, India

Gajra B
Ramanbhai Patel College of Pharmacy, Charotar University of Science and Technology, CHARUSAT Campus, Changa – 388421, Anand, Gujarat, India
Walmart Pharmacy, 3600 Majormeckanzy drive, Vaughan, Ontario, Canada

Parikh RH
Ramanbhai Patel College of Pharmacy, Charotar University of Science and Technology, CHARUSAT Campus, Changa – 388421, Anand, Gujarat, India

Gayatri Patel
Ramanbhai Patel College of Pharmacy, Charotar University of Science and Technology, CHARUSAT Campus, Changa – 388421, Anand, Gujarat, India

Appu Manikandan
Department of Plant Science, Bharathidasan University, Tiruchirappalli, Tamil Nadu, India

Muthukrishnan Sathiyabama
Department of Plant Science, Bharathidasan University, Tiruchirappalli, Tamil Nadu, India

Cintia Rodrigues Maruyama
Laboratory of Biotechnology, University of Sorocaba, Rodovia Raposo Tavares, km 92, 18023-000, Sorocaba, Brazil
Department of Environmental Engineering, São Paulo State University (UNESP), Sorocaba, Brazil
Federal University of São Carlos, Sorocaba campus, Rodovia João Leme dos Santos, km 110, 18052-780, Sorocaba, Brazil

Mariana Guilger
Laboratory of Biotechnology, University of Sorocaba, Rodovia Raposo Tavares, km 92, 18023-000, Sorocaba, Brazil
Federal University of São Carlos, Sorocaba campus, Rodovia João Leme dos Santos, km 110, 18052-780, Sorocaba, Brazil

Mônica Pascoli
Laboratory of Biotechnology, University of Sorocaba, Rodovia Raposo Tavares, km 92, 18023-000, Sorocaba, Brazil
Department of Environmental Engineering, São Paulo State University (UNESP), Sorocaba, Brazil
Federal University of São Carlos, Sorocaba campus, Rodovia João Leme dos Santos, km 110, 18052-780, Sorocaba, Brazil

Natalia Bileshy-José
Laboratory of Biotechnology, University of Sorocaba, Rodovia Raposo Tavares, km 92, 18023-000, Sorocaba, Brazil

P.C. Abhilash
Institute of Environment & Sustainable Development, Banaras Hindu University, Varanasi, India

Leonardo Fernandes Fraceto
Department of Environmental Engineering, São Paulo State University (UNESP), Sorocaba, Brazil

Renata de Lima
Laboratory of Biotechnology, University of Sorocaba, Rodovia Raposo Tavares, km 92, 18023-000, Sorocaba, Brazil
Federal University of São Carlos, Sorocaba campus, Rodovia João Leme dos Santos, km 110, 18052-780, Sorocaba, Brazil

Alireza Samzadeh-Kermani
Department of Chemistry, Faculty of Science, University of Zabol, Zabol, Iran

Negin Esfandiary
Department of Chemistry, Faculty of Science, University of Zabol, Zabol, Iran

Umesh K. Parida
School of Biotechnology, KIIT University, Bhubaneswar, India

Niranjan Rout
Acharya Harihar Regional Cancer Center, Cuttack, India

Birendra K. Bindhani
School of Biotechnology, KIIT University, Bhubaneswar India

Dongzhi Hou
College of Pharmacy, Guangdong Pharmaceutical University, Guangzhou, China

Ruyi Gui
College of Pharmacy, Guangdong Pharmaceutical University, Guangzhou, China

Sheng Hu
College of Pharmacy, Guangdong Pharmaceutical University, Guangzhou, China

Yi Huang
College of Pharmacy, Guangdong Pharmaceutical University, Guangzhou, China

Zuyong Feng
School of Physics and Optoelectronic Engineering, Guangdong University of Technology, Guangzhou, China

Qineng Ping
College of Pharmacy, China Pharmaceutical University, Nanjing, China

PREFACE

The term "nanoparticles and their application" is a broad implication and discussion point for understanding by many experts. But, the question arises that do we have to restrict ourselves to the field of nanomaterials, which consists of building nano-metric structures made up of a limited number of atoms? Nanoparticle based miniaturized world which is largely dominated by application in drug delivery and pharmaceuticals and in which the dimensions of its patent based devices are smaller than a hundredth of a nanometer? How do we approach concepts which at first glance seem to be very different, i.e. the link between volume and surface of the different aggregates, the functionality of macromolecules and the complexity brought about by our biomedical system? In fact, all different scientific disciplines, including every single sector (such as nanomaterials, micro and nano-machines, micro and nano-electronics), have their own paradigm. This is why innovations and industrial developments are profoundly different. However, these fields are strongly interlinked. It is therefore necessary to make our studies more interdisciplinary in order to enable us to understand the nanoparticles based biomedical advantage.

Taking this idea as a basis for our work, we would like to introduce chitosan nanoparticles in the broadest scale possible by showing their optical/chemical characterization with their applicability. We will cover different method of preparation and applications in the chapters to come. This is not a straight textbook; those are easily accessible in libraries or on the Internet. The following chapters will both provoke reflection and provide the reader with a better understanding of the subject. This is a guided tour of the different form of chitosan nanoparticles which we hope will arouse the reader's curiosity so that they will engage more profoundly with the subject.

In many different fields we can observe a tidal wave of new products which are directly linked to nano-sciences and their specific form as chitosan nanoparticles. Therefore, the basic ideas will be introduced and a brief outline will be given in Chapter 1. Yet, it remains difficult to understand how so many complex domains work together.

A basic introduction about synthesis and optimization of chitosan nanoparticles for their potential applications in nano-medicine and biomedical engineering will be given in Chapter 2. In Chapter 3 the synthesis of the tools needed to explore the nano-world will be explained as reactions proceed as micro-droplets reactions. Ionic gelation methods for synthesis in chitosan nanoparticles are covered in Chapter 4, where a new thiolated method has been developed. Chapter 5 will introduce the preparation of chitosan nanoparticles by TPP ionic gelation combined with spray drying, and the antibacterial activity of chitosan nanoparticles and a chitosan nanoparticle–amoxicillin complex.

Chapter 6, shows the understanding the mechanism of ionic gelation for synthesis of chitosan nanoparticles using qualitative techniques. In Chapter 7 we will deal with the a qualitative review on the impact of chitosan composites and chitosan nanoparticle composites on various drug delivery systems. In chapter 8 and 9, we have discussed the modified chitosan moieties. N, N, N- Trimethyl has been used for modifying chitosan and in chapter 9, Ganciclovir loaded chitosan nanoparticles has been used for drug delivery applications. In chapter 10, green synthesis of copper-chitosan nanoparticles has been explained for its antibacterial activity. In last chapter 11, nanoparticles based on chitosan have been developed as carriers for the combined herbicides imazapic and imazapyr. A large part of our discussion takes place in a virtual world, which in turn enables us to understand the nanoworld. Our future depends on how we use new discoveries and what risks they bring upon humanity and our natural environment. The ethical implications of this must therefore be discussed.

Chapters 12-14 bring together research studies and papers, which illustrate some of the developments and applications in the field of chitosan nanoparticles and also give a way forward for future research and application.

Editor

SECTION I

1

CHAPTER

INTRODUCTION ABOUT CHITOSAN NANOPARTICLES

Dr. Abhishek Gupta

SHORT DISCUSSION

Chitosan is the second most abundant polysaccharide next to cellulose on the earth. It is a linear polymer composed of β-(1–4)-2-amino-2-deoxy-D-glucopyranose units. It is white, hard, inelastic and nitrogenous polysaccharide with great potential for wide range of uses due to its biodegradability, biocompatibility, antibacterial activity, non-toxicity and versatile chemical and physical properties. Pharmaceuticals industry, paper production, textiles, wastewater treatment, biotechnology, cosmetics, food processing and agriculture are few of the many areas where chitosan is useful. The presence of primary amine groups makes chitosan an excellent cell transfectant. Like polyethylene amine, chitosan exhibits a "proton sponge" effect, which refers to the swelling behavior of the polymer on encountering an acidic pH inside the cell's endosome, making it an efficient carrier for therapeutic molecules. Chitosan based nanomaterial have superior physical and chemical properties such as high surface area, porosity, tensile strength, conductivity, photo-luminescent as well as increased mechanical properties. In the initial part of the introduction, a brief discussion about chitin and chitosan presented.

Chitin and chitosan

Chitin and chitosan have been investigated thoroughly for different applications, including applications in the especially biological area. Chitin is the most abundant nitrogen-bearing organic compound found in nature and is the second most available polysaccharide next to cellulose. The major sources of chitin production are the cuticles of crustaceans, especially shells of shrimp and crab and other byproducts of the food industry, and the exoskeletons of cephalopods. Chitin is extracted by demineralization and deproteinization of crustacean shells using highly concentrated solutions of sodium hydroxide under high temperature. Chitosan, the amino polysaccharide copolymer of 1,-4 D-glucosamine and N-acetyl glucosamine (Figure 1-2),[1] is derived from chitin by alkaline [2, 3] or enzymatic deacetylation.[4] Therefore, chitin and chitosan are essentially the same polymer but with arbitrarily defined degrees of deacetylation (DD). Generally, if the DD is more than 40%, the term chitosan is used. Although chitin and chitosan are absent in mammals, some mammalian enzymes, such as lysozymes, can hydrolyze them. Lysozymes are nonspecific proteolytic enzymes found in concentrations of 4–13 mg/l in serum and 450-1230 mg/ml in tears.[5] Chitosan's biodegradation process is dependent on two main factors; the DD and the distribution of NAG units. This was elucidated in the works of Lee et al. who showed that biodegradation declines sharply when the DD is more than 9 70%,[6] and Aiba et al. who found that chitosans with randomly distributed NAG units are less susceptible to lysozymic degradation than chitosans which have repeated blocks of three consecutive NAG units.[7] In vitro hydrolysis kinetics of chitosan by lysozymes follows the typical Michaelis-Menten kinetics, where km and Vmax are 15 μg/ml and 0.083x10-3 g/L/min, respectively.[8] However, in vivo degradation of chitosan, which in theory parallels the rate of tissue regeneration, is variable. Tomihata et al. reported that only 10% of a chitosan film was degraded after 50 h of exposure to lysozymes and 80% of the dry film weight was retained after 12 week implantation in Wistar rats.[9]

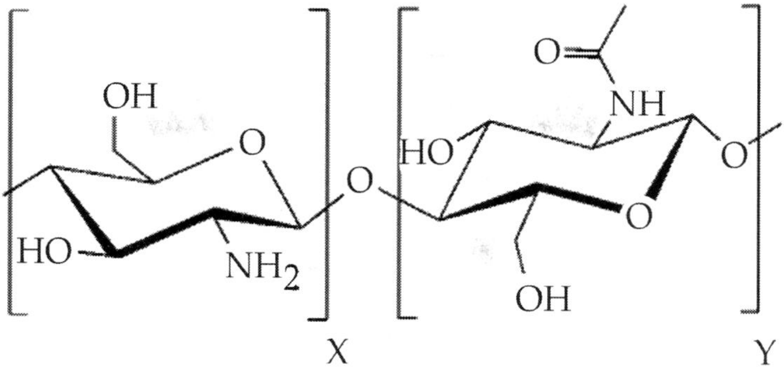

Figure 1.2 Molecular structure of chitosan.

In addition to being biodegradable, chitosan is a biocompatible polymer. This stems from chitosan's distinct structural similarities to the mammalian glycosaminoglycans (GAGs), which are a family of heteropolysaccharides in mammals that are located primarily on the surface of cells and in the extracellular matrix (ECM).[10] These molecules are long unbranched polysaccharides which contain repeating 10 disaccharide units. Each disaccharide unit contains either one of two modified sugars; N-acetyl galactosamine or N-acetyl glucosamine and uronic acid such as glucuronate or iduronate.[11] GAGs have extended conformation that imparts high viscosities and low compressibility properties to the ECM, which renders them suitable for joint lubrication. At the same time, their rigidity provides structural integrity to cells and provides passageways between cells, allowing for cell migration. Chitosan has wide applications in medical fields, such as wound dressing, hypocholesterolemic agents, blood anticoagulant, antithrombogenic and drug delivery systems, in addition to other fields such as waste-water treatment, food and feed additives, wound-healing materials, cosmetic preparations and textile, paper and film technologies.[12,13] Chitosan has many advantages for developing micro/nanoparticles, which can be used for preparing scaffolds and for gene and drug delivery. Chitosan's solubility in aqueous acidic solutions avoids the need for the use of organic solvents when fabricating particulate systems. In addition, the free amino groups become protonated at low pH values which allow the formation of ionic cross-linking with multivalent anions. Moreover, chitosan is mucoadhesive and also provides the ability to sustain the release of active agents such as transforming growth factor-β1.[14]

Chitosan has a significant potential for tissue regeneration purposes because of its biocompatibility, biodegradability and bioresorbability, in addition to its reactivity, having both reactive amino and hydroxyl groups that can be chemically modified. Also chitosan is amenable to molding into porous scaffolds with controllable characteristics. Moreover, chitosan surfaces support the attachment and the subsequent proliferation and growth of different types of cells, which have been attributed to the high cationic charge density of chitosan.[15] Also, chitosan promotes the production of transforming growth factor-beta 1 (TGF-β1) and platelet-derived growth factor (PDGF)[16] and maintains the chondrogenic phenotype.[17] For instance, Lahiji et al. and others showed that 11 chondrocytes seeded on chitosan film surfaces exhibited spherical morphology and expressed aggrecans and type II collagens.[18,19] Bone and cartilage regeneration therapy is an area where chitosan has been shown to be an attractive candidate. For example, chitosan has been investigated as a support matrix for bones [20] and as a biocompatible coating for orthopedic and craniofacial implant devices using palatal mesenchymal cells.[21] The large interest in chitosan for bone regeneration is due to the similarity of chitosan to GAGs and its ability to interact with common connective tissue components such as collagen to form insoluble complexes. The two most important factors that determine the physicochemical properties, and consequently the specific applications of chitosan, are chitosan's degree of deacetylation (DD) and molecular weight (MW). DD, the percentage of the deacetylated units (glucosamine monomers) in chitosan's chains, affects the chemical, physical and biological properties of chitosan, such as the tensile strength of the films, the ability to chelate metal ions and the immunoadjuvant activity. DD also affects the intrinsic pKa (pKo) of chitosan, leading to the alteration of chitosan's solubility in dilute acidic solutions. Generally, DDs $\geq 40\%$ are soluble in dilute acidic solutions. Protonation of the amino groups of glucosamine units contribute to the disruption of hydrogen bonding, the solvation of cationic sites and then to the solubilization of chitosan when the balance between solvent/polymer and polymer/polymer interactions becomes favorable. When the DD becomes less than 40%, chitosan chains become completely insoluble in water, due to 1) the numerous H-bonds that occur between alcohol, amide and ether functionalities distributed on the repeating units all along the polymer chains and 2) the hydrophobic interactions due to the presence of the methyl groups of the acetamide functions and to the –CH and $-CH_2$ of the glucosidic rings. [22] When the DD is more than 40%, chitosan becomes soluble in acidic solution in a pH dependent manner. Decreasing the DD from 100% to 40% results in increasing the pH in which chitosan is soluble. This is due to 1) the increase in pKo by decreasing the DD and 2) the increase in stiffness of the polymer chain on decreasing DD, which results from the amplified steric hindrance from acetylated residues and the expanded excluded volumes.[22-24] The effect of DD on chitosan use in biological systems appears clearly in its effect on gene delivery properties. Kiang et al. investigated the effect of DD of chitosan on the efficiency of gene transfection and found that DNA binding efficacy decreased with decreasing DD, therefore, requiring an increased positive to negative charge ratio for complete DNA complexation.However, the in vivo and in vitro behaviors were different. Decreasing DD resulted in decreased overall luciferase expression levels in vitro but increased luciferase expression levels in vivo up to two fold. The decreased in vitro transfection efficiency can be due to destabilization of the nanoparticles by the bulky acetyl groups in the polymer chains, but this same destabilization, along with higher degradation, were hypothesized to improve the transgene expression in muscles. Therefore, it can be concluded that cellular uptake and gene transfection of chitosan/DNA complexes can be mediated by the DD by controlling electrostatic interactions with the cell membrane to proceed to efficient transfection.[25] In addition, the effect of the DD on cell adhesion has been studied for different cell lines.[26] It was shown that even as little as 10% difference in the extent of acetylation caused significant modification of chitosan's cell attachment properties. The higher %DD of chitosan, the higher amount of free amino groups ($-NH_2$), which in turn, can become protonated to form cationic amine groups ($-NH_3^+$) producing positively charged surfaces. This polycationic nature of chitosan is expected to enhance the interactions between the chitosan surface and the negatively charged cells, and therefore, chitosans with higher extent of deacetylation facilitate cell adhesion.[26] However, high adhesion of some cell lines, such as fibroblasts, to the highly deacetylated chitosan films was also shown to affect cell proliferation. [27,28] These results show that DD should be carefully selected in order to obtain optimum cellular adhesion and proliferation as well as film biodegradation in vitro and in vivo. 13 The molecular weight (MW) of chitosan has significant impact on its effectiveness for a variety of applications. This appears in the efficacy of chitosan to accelerate burn healing, coagulate pollutants, lower blood cholesterol levels, enhance drug dissolution, control viscosity or improve crop yields, which were found to be molecular weight dependent.[29] The size of chitosan polymers necessary to achieve this wide range of application varies by 3 to 4 orders of magnitude. For example, it was reported that 9.3 kDa chitosan prevented the growth of E coli, but 2.2 kDa chitosan promoted its growth.[30] Therefore, it is essential to precisely control the molecular weight of chitosan to suit the different applications and outcomes. This can be done using low concentrations of sodium nitrite ($NaNO_2$) to depolymerize chitosan chains in a controlled manner, offering a possible solution to problems regarding consistency of commercially available chitosans.[29, 31] The effect of chitosan's molecular weight on gene delivery systems interplays in the complex formation with DNA. Once the electrostatic interactions between polyelectrolytes occur, high MW chitosans result in more entanglement that leads to stronger complexes. With decreasing MW of the chitosan, chain entanglement contribution to complex formations decreases, yielding weaker complexes. Therefore, low MW chitosans are less efficient in retaining condensed DNA upon dilution and are less capable of protecting the DNA from degradation by DNase and other serum components. Huang et al. found that these factors resulted in lower transfection efficiencies when using low MW chitosans.83 However, by increasing the charge ratio (N/P) of chitosan to DNA, the transfection efficiencies of low MW chitosans complexed with DNA supersede that of high MW.[32] Hence, it appears that a fine balance must be achieved between extracellular DNA protection (better with high MW) versus efficient intracellular unpacking (better with low MW) to obtain high transgene expression of chitosan.

In tissue regeneration, the effect of chitosan's molecular weight is most important on the physicochemical properties of chitosan scaffolds.[33] Using higher molecular weight chitosans lead to higher tensile strength due to stronger inter-chain associations via Hbonding, which also results in a lower solubility. On the other hand, molecular weight has a minimal direct effect on cell attachment; the lower molecular weight chitosans show only slight enhancement in cell attachment.[34] 1.5 Modifications of chitosan Chitosan is the only cationic polysaccharide in nature derived from biomass. Its unique physicochemical and biological properties make it worthy in regard to pharmaceutical and biomedical applications. However, chitosan properties, such as aqueous solubility in neutral to basic media and mechanical characteristics, can be enhanced via chemical modifications, physical interactions (inorganic composites and polyelectrolyte complexes) and other miscellaneous methods (such as using avidin-biotin interaction). Some of the trials reported in literature to exploit the unique properties and to realize full potential of this versatile polysaccharide, especially in the gene delivery and tissue engineering fields, are discussed below.

Chemical modifications

Chemical modifications of chitosan have been investigated extensively in literature for improving the physicochemical properties of chitosan. The main goals of modifying chitosan chemically are to provide derivatives that are soluble at neutral and basic pH values, to control hydrophobic, cationic, and anionic properties as well as to attach various functional groups and ligands. Fortunately, chitosan is amenable to chemical modifications due to having of hydroxyl, acetamido and amine functional groups.[35] For that reason, generally speaking, chemical modifications would not change the fundamental skeleton of chitosan and would keep the original physicochemical and biochemical properties while bringing new or improved properties.

PEG-grafted chitosan (chitosan-g-PEG) was synthesized by Jiang et al. using the reaction between methoxy PEG-nitrophenol carbonate and chitosan.[36] The average size of nanoparticles prepared from chitosan/DNA was slightly smaller than that of chitosan-g-PEG/DNA complexes in the preparation media, however, chitosan/DNA complexes were highly susceptible to aggregation in the presence of serum and bile. PEG-grafted chitosans effectively shielded the positively charged chitosan surface when compared to unmodified chitosans and were able to maintain complex sizes in the presence of bile and serum, in addition to affording protection to the complexed DNA against enzymatic degradation. Ultimately, this improved the transfection efficiency in vivo. Chitosan-graft-polyethylenimine (CHI-g-PEI) copolymer was synthesized by imine reaction in two steps. In the first step, periodate-oxidized chitosan was prepared by a modified method described by Vold and Christensen,[37] followed by reacting PEI with periodate-oxidized chitosan for two days.[38] It was found that the addition of chitosan to PEI resulted in significant reduction in cytotoxicity compared to PEI (25 kDa), in three different cell lines as measured by the Cell Titer96® assay. In addition, this polymer resulted in enhanced gene transfer efficiency, even in the presence of serum.[38] Trimethylated chitosans (TMCs) were prepared by Kean et al. in order to produce more permanent cationic charges on chitosan's chains.[39] TMCs were evaluated for the effect of the quaternization on cytotoxicity and transfection efficiency. Trimethylation increased the toxicity of chitosans; however, the toxicity remained far below that of PEI up to the 24 h incubation with cells. On the other hand, both trimethyl oligomers (TMOs) and TMCs complexed with pGL3 luciferase reporter gene yielded appreciable transfection in COS7 and MCF7 cell lines. TMOs showed the highest transfection efficiency at 44% degree of methylation (TMO44), whereas, TMCs showed high transfection efficiency only at two methylation degrees, 57 and 93%. TMO44 yielded 40% of PEI transfection in COS-7 and a 16-fold increase over PEI in MCF-7 cells, while 16 TMCs with degrees of trimethylation of 57% and 93% gave 23 and50-fold increase, respectively, over PEI transfection efficiency in MCF-7 cells.[39]

The toxicities and transfection efficiencies of trimethylated chitosan derivatives were dependent on size and degree of methylation. Galactose molecules were coupled to chitosan chains using lactobionic acid bearing the galactose groups via active ester intermediates using 1-ethyl-3-(3-dimethylaminopropyl)-carbodiimide hydrochloride (EDC). In addition, the prepared polymers were grafted with hydrophilic dextrans, and their structure was confirmed using IR.97 This polymer has the potential to be used as a hepatocyte-targeting gene delivery system since galactose functions as a specific ligand to asialoglycoprotein receptors (ASGRs) and dextran sulfate serves as the hydrophilic ligand.[39] Similarly, galactosylated chitosans (GCs) were tested for hepatocyte's adhesion in comparison to galactosecarrying polystyrene (PS), poly(N-p-vinylbenzyl-4-o-β-D-galactopyranosyl-D- gluconamide) (PVLA), that mediates hepatocyte's adhesion through the galactosespecific interactions between ASGRs of the hepatocytes and galactose residues of the PVLA. [40] The results showed that hepatocyte's adhesion to the GC-coated PS dish was as high as 94.7% after 120 min incubation, which was similar to PVLA-coated PS dish. In contrast, hepatocytes adhesion to unmodified chitosan-coated PS dishes was about 69.1%. Finally, the morphology of hepatocytes was studied in the presence and absence of epidermal growth factor (EGF). It was shown that at low concentrations of GC (0.05 μg/ml), cells adhered to the surface showing extended spreading shapes after 24 h in the absence of EGF whereas this happened after 10 h in the presence of EGF, indicating enhanced spreading in the presence of EGF. Therefore, GC showed potential as a synthetic ECM for liver tissue engineering.[41] Galactose-modified chitosans were also coupled with hydrophilic PEG forming galactosylated chitosan-graft-poly(ethylene glycol) (GCP).[42] The incorporation of a flexible hydrophilic polymer into the gene delivery systems produces steric barriers to interaction with proteins and phagocytes and 17 prolonged plasma circulation time. Turbidity measurements were performed to study the dispersive stability of GCP/DNA complexes. GCP/DNA complexes expressed constant or slightly decreasing turbidity without precipitation, which is an indication of enhanced stability. The grafted PEG chains were suggested to be responsible for protecting the GCP/DNA complexes from self-aggregation and precipitation. Similarly, polyvinylpyrrolidone (PVP) was conjugated as another hydrophilic group onto GC chains. PVP is a nontoxic, water soluble neutral hydrophilic polymer that is used as an additive in pharmaceuticals. GC was chemically conjugated with PVP via the formation of an amide bond between the amino group of GC and the activated terminal carboxyl group of the PVP by N-hydroxysuccinimide (NHS)/EDC dissolved in buffer solution.

The purpose of adding hydrophilic ligands such as PVP was mainly to reduce clearance by the reticuloendothelial systems (RES), and to minimize mononuclear phagocyte system uptake, extending circulation time of these particles in vivo.[43] Moreover, it was found that hydrophilic groups, introduced onto the surface of DNA complexes, prevent them from aggregating and being adsorbed onto plasma proteins, especially albumin, which can stimulate the clearance of DNA vectors from the plasma by RES. The effect of albumin on the surface charge of GC-PVP 10K/DNA complexes were examined and compared to PEI, a commonly used polycationic gene carrier. The zeta potential values of PEI/DNA complexes at charge ratios of 1 and 3 were significantly decreased to negative values even at low concentrations of albumin. On the other hand, GC-PVP 10K/DNA had a constant surface charge regardless of the amount of albumin added.

Mannose receptors have an important role in endocytosis and are expressed in high levels in antigen presenting cells, such as dendritic cells (DCs) which are essential for activating immune response. Gene modification of DCs is of particular interest for immunotherapy of diseases where the immune system has failed or is abnormally regulated, such as in cancer or autoimmune diseases. In these cases increasing the 18 synthesis of immunomodulatory cytokines such as IL-12 in tumor tissues can allow better T-cell activation.[44] Mannosylated chitosans (MCs) were prepared by reacting water soluble chitosans with mannopyranosyl phenylisothiocyanate in dimethyl sulfoxide (DMSO).[44] MC/DNA complexes were induced to self-assemble by mixing the plasmid DNA with MC in phosphate buffer at pH 7.4. The transfection results of these complexes clearly show that MC-mediated pGL3 DNA transfection was mannose receptor– dependent, as confirmed by the ability of mannose to block the binding of complexes onto Raw 264.7 macrophage cells expressing moderate mannose receptors in a concentration-dependent manner.[45] In addition, the levels of plasmid encoding murine IL-12 (pmIL-12), as well as murine IFN-γ, were significantly increased by both MC/pmIL-12 and chitosan/pmIL-12 complexes compared with naked mIL-12, but MC-mediated cytokine production was distinctively higher than that of chitosan's. Similarly, MC showed promising results in transferring genes to the mouse peritoneal macrophages.[46] The water soluble carbodiimide (EDC) was reacted with the carboxyl groups of deoxycholic acid to form active ester intermediates. These intermediates were reacted with the primary amine of chitosan to form an amide bond, releasing isourea as a byproduct, and the formation of the amide bonds was confirmed by FTIR.[47] Hydrophobically modified chitosans by deoxycholic acid (DAMCs) provide colloidally stable self-aggregates in aqueous media. The prepared self-aggregates had small sizes (mean diameter of ca. 160 nm) with a unimodal size distribution, and were able to form charge complexes when mixed with plasmid DNA.[48, 49] In addition, it was found that the optimum condition for cell transfection could be controlled using DAMC self-aggregates with different physicochemical properties, including size and structure.

In a similar fashion, highly purified chitosan oligosaccharides

(COSs), prepared by ultrafiltration methods, were chemically modified with hydrophobic moieties of deoxycholic acid.[50] The hydrophobized COSs (COSDs) formed core-shell type 19 nanoparticles in aqueous media due to their amphiphilic character. Compared to unmodified COSs, the COSDs nanoparticles showed greater potential as gene carriers by having more efficient DNA condensation and protection. The COSDs, especially COS3D25 (Mn = 3 kDa, degree of substitution = 5.4%), mediated high levels of gene transfection in HEK293 cells compared to that of poly(L-lysine) (PLL), in the absence and presence of FBS. Chitosan oligosaccharides were also modified chemically with hydrophobic cholesterol groups. Cholesterol-modified chitosan oligosaccharides (COSs) were fractionized according to molecular weight using ultrafiltration technique.[51] COSs were dissolved in DMSO/H_2O mixture and reacted with cholesteryl chloroformate to form hydrophobized COS.

Due to their amphiphilic characteristics, modified COSs formed core shell nanoparticles with low critical aggregation concentration in aqueous milieu. COS6C5 (Mn= 6 kDa, containing 5% of cholesteryl chloroformate) showed the best DNA condensation and the highest transfection efficiency compared to the unmodified chitosans.[51] Alkylated chitosans (ACSs) were prepared by modifying chitosan with alkyl bromide.[52] It was found that ACSs self-aggregate in acetic acid solution, whereas 99% deacetylated chitosans do not, due to strong electrostatic repulsion. The transfection efficiency of plasmid-encoding chloramphenicol acetyltransferase (CAT) mediated by chitosan and ACS into C_2C_{12} cells, a mouse skeletal muscle cell line, was dependent on the hydrophobicity of chitosan. Increasing the length of the added alkyl side chain, up to 8 carbons, caused the transfection efficiency to increase, after which it leveled off. It was suggested that higher transfection efficiencies of ACS was due to easier unpacking of DNA from ACS carriers inside cells in addition to enhancing entry into cells facilitated by hydrophobic interactions.

Primary amines of glycol chitosans were modified with 5β-cholanic acid to prepare hydrophobically modified glycol chitosan (HGC).[53] DNA was complexed with 20 cetyltrimethylammonium bromide (CTAB) at a charge ratio of 1, which resulted in the highest DNA stability. This was followed by encapsulating the DNA in the HGC nanoparticles by hydrophobic interactions between 5β-cholanic acid moieties and the hydrophobized DNA. Increasing the HGC/GC ratio caused the encapsulation efficiency of DNA to increase slowly and the particle sizes to decrease. This was attributed to the hydrophobic interactions between 5β-cholanic acids of HGC and CTAB/DNA complexes. HGC with high degrees of 5β-cholanic acids substitution were found to have lower critical micelle concentrations (CMC) and to have the ability to be formulated in smaller particles.[53,54]

Amphiphilic grafted copolymers based on CSO were prepared by conjugating stearic acid (SA) molecules to chitosan chains to form hydrophobic segments using EDC chemistry.[55] CSO-SA micelles formed spontaneously upon dispersion in distilled water with the help of ultrasonication and were found to have CMC value of 0.035 mg/ml as determined by fluorometry with pyrene. The CSO–SA micelles were used to condense plasmid DNA and the CSO–SA/DNA complex nanoparticles with N/P ratios from 0.25 to 58 were tested for DNA stability, cytotoxicity and in vitro transfection efficiency.[56] In vitro transfection testing showed higher transfection efficiencies of CSO–SA micelles in A549 cells compared to that of CSO. Moreover, the transfection efficiencies of these micelles were comparable to Lipofectamine™ 2000 in serum free media, but greater than this known transfection agent in the presence of 10% fetal bovine serum. On the other hand, the cytotoxicity of CSO–SA was significantly lower than that of Lipofectamine™ 2000 (The IC50 of CSO–SA and Lipofectamine™ 2000 were found to be 543 μg/ml and 6 μg/ml, respectively). The low cytotoxicity of the studied CSO–SA micellar vectors in conjunction with superior DNA condensation capacity are critical improvements for nonviral gene transfer.

Poly (chitosan-g-DL-lactic acid) (PCLA) was investigated for tissue engineering applications. PCLA was synthesized by dissolving chitosan in an aqueous solution of 21 DL-lactic acid, and then the polymer was left on a Teflon dish at 85°C to shape into a membrane during the ongoing polymerization for 5 h.[57] During the synthesis of the PCLA copolymer, chitosan amino lactate salts were formed first through the protonation of chitosan. This was followed by establishing the real linkages between chitosan and lactic acid by dehydrating the resultant lactate salts while the polycondensation of lactic acid was happening simultaneously.[56] Fibrous mesh scaffolds were successfully fabricated by using PCLA with an improved wet-spinning technique. Several key processing conditions, such as the concentration of PCLA dopes, the nitrogen pressure applied to the dopes, and the concentration of coagulant were used to modulate the morphologies of the fabricated filaments and the structure of the scaffolds. PCLA mesh scaffolds showed well-defined tensile and compressive mechanical properties similar to chitosan scaffolds in their dry states. After being hydrated, greatly improved tensile and compressive characteristics as well as enhanced dimensional stability in PCLA mesh scaffolds were obtained, in comparison to wet chitosan mesh scaffolds.

Chitosan composites

Chitin, the precursor of chitosan, exists as a natural composite with minerals and proteins in invertebrates and functions as essential component for preserving the structural integrity of the shells of arthropods.[57] Also, blends or composites of chitosan with other materials have been widely investigated in literature. Such blends have resulted in a range of modifications of chitosan properties and applications.[58] Chitosan composites can be classified into two main categories; chitosan–inorganic material composites and chitosan polyanion complex composites. Various inorganic materials, including metals, have been incorporated into chitosan composites which resulted in unusual and interesting biochemical properties.[59] The collective characteristics of these admixtures of inorganic compounds with chitosan 22 have been applied in several appealing biotechnological applications. Hydroxyapatite, for example, was extensively studied in combination with chitosan as a bioactive inorganic material to further enhance bone regenerative efficacy and osteoconductivity.[60] These materials have been applied in bone tissue recovery, such as bone-filling materials, and in drug delivery systems intended to accelerate reconstitution of damaged bones.[61,62]

The focus of this study is on the second type of chitosan composites, polyelectrolyte complexes (PECs). Complexes between macromolecules can be formed by several types of intermolecular interactions. Some of these forces include van der Waals interactions, hydrogen bonding, coordination forces, covalent bonding and polyelectrolyte interactions to form PECs. Generally, combined interpolymer forces contributes to the stability of the macromolecular complexes.[63] PECs formation is a simple but very attractive principle of particle formation with high practical potential. PECs are formed as a result of condensation between oppositely charged polyions that are stabilized by intermolecular ionic bonds, and their main driving force for the reaction is the concomitant release of corresponding counterions, increasing the entropy of the system.[64] PECs can be generally categorized into two types; water-soluble and aggregated PECs. Water-soluble PECs are formed as a result of non-stoichiometric mixing of weakly charged polyelectrolytes that varies in molecular dimensions, whereas water insoluble and highly aggregated PECs are formed from highly charged and/or high molecular weight polyelectrolytes.[65] One of the more widely used polycations for the fabrication of PECs is chitosan. Chitosan is considered the only available cation from biomass. This cationic property of chitosan allows it to form non-covalent complexes with other negatively charged polyelectrolytes. Examples of polyanions used to form PECs with chitosan are DNA,[66] carboxymethyl cellulose, [67] gum kondagogu, [68] sodium alginate, [69-70], acrylamido glycolic acid, [71] poly(acrylic acid), [72] pectin, [73] hyaluronic acid [74] and sodium dextran sulfate. [75] The complexation process of chitosan with polyanions depends on many factors, including the degree of deacetylation of chitosan, and consequently the dissociation constant (pKa), the molecular weight of chitosan, the degree of polydispersity, chitosan's chain conformation, polymer's concentrations and ratios, charge density and the distribution of amine to acetylated groups along the polymer chains. In addition, the complexation environment, such as solution pH, temperature and ionic strength plays an important role. [63-76], For example, changing the ionic strength by adding salt to the chitosan solution will partially inhibit the repulsion forces between charged groups throughout chitosan chains, which decreases the rigidity of the chains. The substantial role of chitosan's degree of deacetylation on the complexation process appears in the effect on chain conformation. For instance, at a degree of deacetylation of more than 80%, chain extension happens due to the predomination of electrostatic repulsion between protonated amino groups. At degrees of deacetylation between 50% and 80%, the effect of acetyl groups becomes dominant, causing stronger intermolecular hydrogen bonds and increased steric hindrance of acetamide groups compared to amino groups. This limits the flexibility of chitosan chains and increases their rigidity. Decreasing the degrees of deacetylation to values less than 50% causes the polymer to start aggregating, which increases the local concentrations of polymeric segments. [76]

In addition to forming films, PECs between chitosans and other negatively charged molecules have been utilized to form nanoparticles in a process called ionic gelation. Other methods commonly used for the preparation of chitosan nanoparticles include spray drying, emulsion cross-linking and complex coacervation. However, ionic gelation technique gained the most interest because of its dependence on the electrostatic interaction mechanism, which 1) obviates the use of the commonly harsh conditions associated with chemical cross-linking and residual effects of organic solvents and 2) 24 provides a reversible process that can be optimized and controlled.[14,77] Polyanions such as sodium tripolyphosphate, sodium polyphosphate and sodium sulfate have been investigated for cross-linking chitosan chains. However, chitosan nanoparticles fabricated using ionic gelation have stability issues. For examples, chitosan gel nanoparticles formed by ionic interactions of chitosan chains with tripolyphosphate were found to be extremely labile to environmental conditions such as pH and salt content.[78] Changes in pH values affect the swelling of particles; the more acidic the solution, the larger the diameter of chitosan particles due to the intra-molecular electrostatic repulsions. Also, particle aggregation has been observed in basic pH solutions. Salt content, on the other hand, caused particle disintegration due to swelling of the particle as a result of the difference in osmotic pressure. Gelatin, an example of a polyanion used for the fabrication of chitosan composites, was incorporated with chitosan to fabricate tissue regeneration scaffolds.[79] In the presence of gelatin, mechanical properties of chitosan membranes were significantly affected. Gelatin caused the stiffness of chitosan membranes to decrease in dry conditions but to increase significantly in wet conditions, despite the fact that gelatin possesses very low stiffness relative to chitosan in wet conditions.[79] In addition, the presence of gelatin enhanced the degradation rate and maintained the dimensions of the membranes in the presence of lysozymes.136 In vivo study was performed by isolating autologous chondrocytes from pig's auricular cartilage and seeding them onto chitosangelatin scaffolds. This scaffold was successfully engineered into elastic cartilages that have acquired not only normal histological and biochemical, but also mechanical properties after 16 weeks of implantation.[80] γ-Poly (glutamic acid) (γ-PGA), a hydrophilic and biodegradable polymer, was used to produce γ-PGA/chitosan composite biomaterials in order to enhance the hydrophilicity and cytocompatibility of chitosan-based matrices.[81] To prepare this composite, chitosan powder was uniformly dispersed in γ-PGA solution in water, 25 followed by the addition of acetic acid which caused the chitosan powder to immediately dissolve, thus avoiding aggregation caused by γ-PGA/chitosan polyion complexes and forming a homogeneous solution. Dense films were prepared by drying in the oven, whereas porous films were prepared by freeze-gelation. It was found that the porous matrices have an interconnected pore structure with a pore size of 30 to 100 μm. The addition of γ-PGA increased the maximum load (strength) of the composite matrices. Also hydrophilicity and serum protein adsorption properties of the composite matrices were significantly enhanced. Cytocompatibility of γ-PGA/chitosan composites was enhanced as shown by increased ROS cell attachment and proliferation when compared to chitosan matrices. This makes γ-PGA/chitosan composite matrix a very promising biomaterial for tissue engineering applications.[81] Collagen/chitosan scaffolds were prepared in order to overcome the fast biodegradation rate and the low mechanical strength of untreated collagen scaffolds for skin regeneration purposes. The scaffolds were fabricated by mixing the two polymers followed by lyophilization in cold ethanol. Finally, the two polymers were cross-linked with glutaraldehyde. Confocal laser scanning microscopy (CLSM) images of FITC-chitosan and rhodamine-collagen confirmed that the scaffold was composed of chitosan and collagen filaments which were evenly dispersed throughout the scaffold. Cross-linking with glutaraldehyde resulted in larger pore sizes and in less biodegradability compared to collagen alone, in addition to higher tensile strength compared to uncross-linked chitosans. However, the wettability also decreased significantly, as shown by the swelling test of the different scaffolds.[82] It appears that chitosan/collagen composites are promising scaffold materials for periodontal tissue engineering, as well as for dermal equivalents.[82] Chitosan-alginate composites were studied for potential uses as scaffolding materials, especially for cartilage repair and regeneration.[83] Chitosan and alginate solutions were prepared separately by dissolving chitosan and sodium alginate in 26 1N acetic acid and NaOH, respectively, and then mixing the two solutions together. The lyophilized discs were then cross-linked by immersion in 1% w/v CaCl2 solution. The cells increased consistently on both chitosan and chitosan-alginate scaffolds over cell culture time, and proliferated faster on the chitosan-alginate scaffold than on the chitosan scaffold. Furthermore, the cell viability assay revealed more live cells on chitosanalginate than on pure chitosan. These results are due to the known property of alginate to promote cell expansion; which augmented chitosan's excellent cell adhesive property due to its cationic nature. Thus, chitosan-alginate hybrid materials may have combined favorable properties of chitosan and alginate in terms of cellular attachment and proliferation.140 In addition, chitosan-alginate composites in the form of nanoparticles were prepared in mild aqueous conditions through ionic gelation, driven by the electrostatic interactions between the two species.[84] The prepared nanoparticles had small size and narrow distribution. Particles had a mean Z-average diameter of ca. 157 nm and a zeta potential of 32 mV. Chitosan-alginate nanoparticle-mediated transfection of HEK293T cells resulted in transfection levels as high as achieved with Lipofectamine™ after 48 h. Transfection efficiency of complexes prepared at a 5:1 N/P ratio was higher than with chitosan nanoparticles or naked plasmid. The improvement in transfection efficiency was explained by the presence of alginate which reduces the interaction strength between chitosan and DNA but does not prevent it, helping the dissociation of the complexes inside the cells.

One of the macromolecules that shows a great practical potential for complexing with chitosan for forming composites is dextran sulfate (Figure 1-3).[75, 85] Dextran sulfate, a biodegradable polymer similar to heparin, contains approximately 17% sulfur, which is equivalent to approximately 2.3 sulfate groups per glucosyl residue. Dextran sulfate is derived by esterification of dextrans using sulfuric acid. Dextran is a polymer of anhydroglucose and is composed mainly of alpha-D-(1-6) linkages and a small percentage of (1-3) linkages that account for the branching of dextran. Most 27 commercially available dextrans are obtained from Leuconostoc mesenteroides. Some of the dextran sulfate applications include lipoprotein separation, accelerating hybridization, releasing DNA from DNA-histone complexes and inhibiting ribonucleases. Dextran sulfate showed potential to be used as antiviral agent and for enzyme and protein stabilization, as well as in cosmetic preparations.

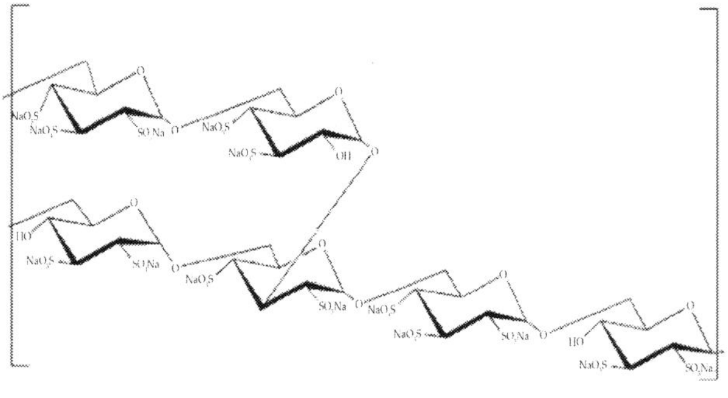

Figure 1.3 Molecular structure of dextran sulfate.

The mechanism of formation of colloidal PECs from chitosan and dextran sulfate depends on which species is in excess. When dextran sulfate is in excess, Drogoz et al. proposed that the two polymers interact to form primary complexes, which rearrange to form colloidal species.[64] In this case, the mechanism of polymer interaction depends on chitosan molecular weight. For low molecular weight chitosans, a residue-to-residue mechanism was proposed, which results from the guest-host model introduced by 28 Kabanov.[86] This requires large difference of molecular weights between the two polymers and a stiff guest polymer. On the other hand, using high molecular weight chitosans and when chitosan is in excess, charge-to-charge neutralization predominates. Chitosan-dextran sulfate PECs have been studied for a wide range of applications, mainly for peptide delivery, such as anti-angiogenesis,[87] hexapeptide dalargin[88], repifermin (R)[89], HIV-1 p24 antigen [90] and insulin.[91] In addition, chitosan-dextran sulfate PECs have been studied for delivery of small molecules such as amphotericin B.[92]

Modifications via avidin-biotin

Biotin, hexahydro-2-oxo-1H-thieno[3,4-d]imidazole-4-pentanoic acid or vitamin H, is a naturally occurring vitamin found in all living cells. The tissues with the highest amounts of biotin are the liver, kidney and pancreas. Cancerous tumors have more biotin than normal tissue.[93] Biotin is involved as an enzyme cofactor in a variety of carboxylase, decarboxylase and transcarboxylaze reactions,[94-96] and is known to have a strong affinity to a number of proteins, such as avidin and streptavidin. Avidin is a tetrameric glycoprotein which was originally isolated from chicken egg white and is found in the tissues of birds. Avidin has four identical subunits, which consist of 128 amino acids each.

The reported molecular weight of avidin ranges from 66-69 kDa depending on the method of analysis. Avidin is soluble in aqueous solutions and stable over a wide pH and temperature range. Early work involving avidin-biotin chemistry centered on biotin's function as a vitamin, yet in 1927, rats fed large quantities of egg white developed dermatitis, indicating malnutrition. Vitamin H, structurally identified as biotin in 1940, prevented this dermatitis. The malnutrition induced by avidin was eventually attributed to the depletion of biotin.[97]

The avidin-biotin interaction is the strongest known non-covalent, biological interaction between a protein and a ligand (Kd for avidin-biotin complex =10-15M-1). The binding capacity of avidin is 4 moles biotin to 1 mole avidin. Bond formation is very rapid and once formed is unaffected by wide extremes of pH, temperature, organic solvents and other denaturing agents. The avidin-biotin complex is also resistant to enzymatic proteolysis. Avidin-biotin chemistry is popular for a broad range of technologies, such as diagnostic applications, biosensorin[99] and affinity based separation.[98] Also, avidinbiotin technology has proved to be a versatile tool for a variety of pharmaceutical applications such as targeted drug delivery.[100] As an example of applications of avidin-biotin technology, novel biotinylated nanotemplated degradable hydrogels that integrate cell interactivity into their surface were prepared.[101] These kinds of modifications are critical in the development of biomaterials for drug delivery and tissue engineering applications. Biotin modification has been applied to several polymers, including poly(styrene-co-N-acryloxysuccinimide),[102] poly(ethylene glycol) (PEG)-ylated polyethyleneimine,[103] PEG-poly[lactic-co-(glycolic acid)] (PLGA)[104] and poly(lactic acid)-PEG.[105] This method has been studied only in a limited way for natural polymers such as chitosan. For instance, avidin-biotin interaction has been applied to prepare chitosan supports for protein electrodeposition by Shi et al.[106] Chitosans modified using this technique have potential for a vast range of applications in different biomedical fields. Nonetheless, utilizing the numerous advantages of this polymer using this technique necessitates studying chitosan biotinylation reaction, ligand attachment conditions and fabrication methods.

REFERENCES

1. Chung, T.W. et al. (2003) Growth of human endothelial cells on different concentrations of Gly-Arg-Gly-Asp grafted chitosan surface. *Artificial Organs* 27 (2), 155-161.
2. Kurita, K. et al. (1977) Studies on chitin, .4. Evidence for formation of block and random copolymers of N-acetyl-D-glucosamine and D-glucosamine by heterogeneous and homogeneous hydrolyses *Macromolecular Chemistry and Physics* 178 3197.
3. Varum, K.M. et al. (1991) High-field NMR-spectroscopy of partially Ndeacetylated chitins (chitosans) .3. C-13-NMR studies of the acetylation sequences in partially N-deacetylated chitins (chitosans). *Carbohydrate Research* 217, 19-27.
4. Araki, Y. and Ito, E. (1975) Pathway of chitosan formation in Mucor-Rouxii - enzymatic deacetylation of chitin. *European Journal of Biochemistry* 55, 71-78.
5. Nordtveit, R.J. et al. (1996) Degradation of partially N-acetylated chitosans with hen egg white and human lysozyme. *Carbohydrate Polymers* 29 (2), 163-167.
6. Lee, K.Y. et al. (1995) Blood compatibility and biodegradability of partially Nacylated chitosan derivatives. *Biomaterials* 16 (16), 1211-1216.
7. Aiba, S. (1992) Studies on chitosan .4. lysozymic hydrolysis of partially Nacetylated chitosans. *International Journal of Biological Macromolecules* 14 (4), 225-228.
8. Pangburn, S.H. et al. (1982) Lysozyme degradation of partially deacetylated chitin, its films and hydrogels. *Biomaterials* 3 (2), 105-108.
9. Tomihata, K. and Ikada, Y. (1997) In vitro and in vivo degradation of films of chitin and its deacetylated derivatives. *Biomaterials* 18 (7), 567-575.
10. Griffon, D.J. et al. (2006) Chitosan scaffolds: Interconnective pore

size and cartilage engineering. *Acta Biomaterialia* 2 (3), 313-320.

11. Seyrek, E. and Dubin, P. (2010) Glycosaminoglycans as polyelectrolytes. *Advances in Colloid and Interface Science* 158 (1-2), 119-129.
12. Hirano, S. (1996) Chitin biotechnology applications. *Biotechnology Annual Review* 2, 237-258.
13. Struszczyk, M.H. (2002) Chitin and chitosan - Part II. Applications of chitosan. *Polimery* 47 (6), 396-403.
14. Agnihotri, S.A. et al. (2004) Recent advances on chitosan-based micro- and nanoparticles in drug delivery. *Journal of Controlled Release* 100 (1), 5-28.
15. Bumgardner, J.D. et al. (2003) Contact angle, protein adsorption and osteoblast precursor cell attachment to chitosan coatings bonded to titanium. *Journal of Biomaterials Science-Polymer Edition* 14 (12), 1401-1409.
16. Ueno, H. et al. (2001) Evaluation effects of chitosan for the extracellular matrix production by fibroblasts and the growth factors production by macrophages. *Biomaterials* 22 (15), 2125-2130.
17. Sechriest, V.F. et al. (1999) GAG-augmented polysaccharide hydrogel: A novel biocompatible and biodegradable material to support chondrogenesis. *Journal of Biomedical Materials Research* 49 (4), 534-541.
18. Lahiji, A. et al. (2000) Chitosan supports the expression of extracellular matrix proteins in human osteoblasts and chondrocytes. *Journal of Biomedical Materials Research Part A* 51 (4), 586-595.
19. Nettles, D.L. et al. (2002) Potential use of chitosan as a cell scaffold material for cartilage tissue engineering. *Tissue Engineering* 8 (6), 1009-1016.
20. Lee, Y.M. et al. (2000) Tissue engineered bone formation using chitosan/tricalcium phosphate sponges. *Journal of Periodontology* 71 (3), 410- 417.
21. Bumgardner, J.D. et al. (2003) Chitosan: potential use as a bioactive coating for orthopaedic and craniofacial/dental implants. *Journal of Biomaterials Science- Polymer Edition* 14 (5), 423-438.
22. Lamarque, G. et al. (2005) Physicochemical behavior of

homogeneous series of acetylated chitosans in aqueous solution: role of various structural parameters. *Biomacromolecules* 6 (1), 131-142.

23. Tsukada, S. and Inoue, Y. (1981) Conformational properties of chitooligosaccharides - titration, optical-rotation, and C-13 NMR-studies of chitooligosaccharides. *Carbohydrate Research* 88 (1), 19-38.
24. Ottoy, M.H. et al. (1996) Compositional heterogeneity of heterogeneously deacetylated chitosans. *Carbohydrate Polymers* 29 (1), 17-24.
25. Kiang, T. et al. (2004) The effect of the degree of chitosan deacetylation on the efficiency of gene transfection. *Biomaterials* 25 (22), 5293-5301.
26. Huang, M. et al. (2004) Uptake and cytotoxicity of chitosan molecules and nanoparticles: Effects of molecular weight and degree of deacetylation. *Pharmaceutical Research* 21 (2), 344-353.
27. Prasitsilp, M. et al. (2000) Cellular responses to chitosan in vitro: The importance of deacetylation. *Journal of Materials Science: Materials in Medicine* 11 (12), 773-778.
28. Chellat, F. et al. (2000) Study of biodegradation behavior of chitosan-xanthan microspheres in simulated physiological media. *Journal of Biomedical Materials Research* 53 (5), 592-599.
29. Allan, G.G. and Peyron, M. (1995) Molecular-weight manipulation of chitosan .1. Kinetics of depolymerization by nitrous-acid. *Carbohydrate Research* 277 (2), 257-272.
30. Tokura, S. et al. (1997) Molecular weight dependent antimicrobial activity by chitosan. *Macromolecular Symposia* 120, 1-9.
31. Allan, G.G. and Peyron, M. (1995) Molecular-weight manipulation of chitosan .2. Prediction and control of extent of depolymerization by nitrous-acid. *Carbohydrate Research* 277 (2), 273-282.
32. Koping-Hoggard, M. et al. (2003) Relationship between the physical shape and the efficiency of oligomeric chitosan as a gene delivery system in vitro and in vivo. *Journal of Gene Medicine* 5 (2), 130-141.
33. Zhang, H. and Neau, S.H. (2001) In vitro degradation of chitosan by

a commercial enzyme preparation: effect of molecular weight and degree of deacetylation. *Biomaterials* 22 (12), 1653-1658.

34. Dhulipala, V.C. et al. (2005) Secalonic acid D blocks embryonic palatal mesenchymal cell-cycle by altering the activity of CDK2 and the expression of p2l and cyclin E. *Birth Defects Research Part B: Developmental and Reproductive Toxicology* 74 (3), 233-242.
35. Mourya, V.K. and Inamdar, N.N. (2008) Chitosan-modifications and applications: Opportunities galore. *Reactive & Functional Polymers* 68 (6), 1013-1051.
36. Jiang, X. et al. (2006) Chitosan-g-PEG/DNA complexes deliver gene to the rat liver via intrabiliary and intraportal infusions. *Journal of Gene Medicine* 8 (4), 477-487.
37. Vold, I.M.N. and Christensen, B.E. (2005) Periodate oxidation of chitosans with different chemical compositions. *Carbohydrate Research* 340 (4), 679-684.
38. Jiang, H.L. et al. (2007) Chitosan-graft-polyethylenimine as a gene carrier. *Journal of Controlled Release* 117 (2), 273-280.
39. Kean, T. et al. (2005) Trimethylated chitosans as non-viral gene delivery vectors: Cytotoxicity and transfection efficiency. *Journal of Controlled Release* 103 (3), 643-653.
40. Park, I.K. et al. (2003) Galactosylated chitosan (GC)-graft-poly(vinyl pyrrolidone) (PVP) as hepatocyte-targeting DNA carrier Preparation and physicochemical characterization of GC-graft-PVP/DNA complex (1). *Journal of Controlled Release* 86 (2-3), 349-359.
41. Park, I.K. et al. (2003) Galactosylated chitosan as a synthetic extracellular matrix for hepatocytes attachment. *Biomaterials* 24 (13), 2331-2337.
42. Park, I.K. et al. (2001) Galactosylated chitosan-graft-poly(ethylene glycol) as hepatocyte-targeting DNA carrier. *Journal of Controlled Release* 76 (3), 349-362.
43. Leroux, J.C. et al. (1996) Biodegradable nanoparticles - From sustained release formulations to improved site specific drug delivery. *Journal of Controlled Release* 39 (2-3), 339-350.
44. Kim, T.H. et al. (2006) Mannosylated chitosan nanoparticle-based cytokine gene therapy suppressed cancer growth in BALB/c mice

bearing CT-26 carcinoma cells. *Molecular Cancer Therapeutics* 5 (7), 1723-1732.

45. Kim, T.H. et al. (2006) Receptor-mediated gene delivery into antigen presenting cells using mannosylated chitosan/DNA nanoparticles. *Journal of Nanoscience and Nanotechnology* 6 (9-10), 2796-2803.
46. Hashimoto, M. et al. (2006) Gene transfer by DNA/mannosylated chitosan complexes into mouse peritoneal macrophages. *Biotechnology Letters* 28 (11), 815-821.
47. Lee, K.Y. et al. (1998) Preparation of chitosan self-aggregates as a gene delivery system. *Journal of Controlled Release* 51 (2-3), 213-220.
48. Kim, Y.H. et al. (2001) Structural characteristics of size-controlled selfaggregates of deoxycholic acid-modified chitosan and their application as a DNA delivery carrier. *Bioconjugate Chemistry* 12 (6), 932-938.
49. Kim, K. et al. (2005) Physicochemical characterizations of self-assembled nanoparticles of glycol chitosan-deoxycholic acid conjugates. *Biomacromolecules* 6 (2), 1154-1158.
50. Chae, S.Y. et al. (2005) Deoxycholic acid-conjugated chitosan oligosaccharide nanoparticles for efficient gene carrier. *Journal of Controlled Release* 109 (1-3), 330-344.
51. Son, S.H. et al. (2004) Preparation of a hydrophobized chitosan oligosaccharide for application as an efficient gene carrier. *Macromolecular Research* 12 (6), 573- 580.
52. Liu, W.G. et al. (2003) N-alkylated chitosan as a potential nonviral vector for gene transfection. *Bioconjugate Chemistry* 14 (4), 782-789.
53. Yoo, H.S. et al. (2005) Self-assembled nanoparticles containing hydrophobically modified glycol chitosan for gene delivery. *Journal of Controlled Release* 103 (1), 235-243.
54. Kwon, S. et al. (2003) Physicochemical characteristics of self-assembled nanoparticles based on glycol chitosan bearing 5 beta-cholanic acid. *Langmuir* 19 (24), 10188-10193.
55. Hu, F.Q. et al. (2006) A novel chitosan oligosaccharide-stearic acid micelles for gene delivery: Properties and in vitro transfection

studies. *International Journal of Pharmaceutics* 315 (1-2), 158-166.

56. Wan, Y. et al. (2008) Preparation and mechanical properties of poly(chitosan-g- DL-lactic acid) fibrous mesh scaffolds. *Polymers for Advanced Technologies* 19 (2), 114-123.
57. Raabe, D. et al. (2007) Preferred crystallographic texture of alpha-chitin as a microscopic and macroscopic design principle of the exoskeleton of the lobster Homarus americanus. *Acta Biomaterialia* 3, 882-895.
58. Kim, I.Y. et al. (2008) Chitosan and its derivatives for tissue engineering applications. *Biotechnology Advances* 26 (1), 1-21.
59. Muzzarelli, C. and Muzzarelli, R.A.A. (2002) Natural and artificial chitosaninorganic composites. *Journal of Inorganic Biochemistry* 92 (2), 89-94.
60. Hein, S. et al. (2008) Chitosan composites for biomedical applications: status, challenges and perspectives. *Materials Science and Technology* 24 (9), 1053-1061.
61. Li, X.N. et al. (2010) Synthesis and Characterization of Core-shell Hydroxyapatite/chitosan Biocomposite Nanospheres. *Journal of Wuhan University of Technology-Materials Science Edition* 25 (2), 252-256.
62. Wilson, O.C. and Hull, J.R. (2008) Surface modification of nanophase hydroxyapatite with chitosan. *Materials Science and Engineering C-Biomimetic and Supramolecular Systems* 28 (3), 434-437.
63. Krayukhina, M.A. et al. (2008) Polyelectrolyte complexes of chitosan: formation, properties, and applications. *Uspekhi Khimii* 77 (9), 854-869.
64. Drogoz, A. et al. (2007) Polyelectrolyte complexes from polysaccharides: Formation and stoichiometry monitoring. *Langmuir* 23 (22), 10950-10958.
65. Dautzenberg, H. et al. (1996) Stoichiometry and structure of polyelectrolyte complex particles in diluted solutions. pp. 1024-1032, Vch Publishers Inc.
66. Alatorre-Meda, M. et al. (2009) DNA-chitosan complexation: A dynamic light scattering study. *Colloids and Surfaces*

A-Physicochemical and Engineering Aspects 339 (1-3), 145-152.

67. Zhao, Q. et al. (2009) Synthesis and characterization of soluble chitosan/sodium carboxymethyl cellulose polyelectrolyte complexes and the pervaporation dehydration of their homogeneous membranes. *Journal of Membrane Science* 333 (1-2), 68-78.
68. Naidu, V.G.M. et al. (2009) Polyelectrolyte complexes of gum kondagogu and chitosan, as diclofenac carriers. *Carbohydrate Polymers* 76 (3), 464-471.
69. Li, X.X. et al. (2009) Characterization and biodegradation of chitosan-alginate polyelectrolyte complexes. *Polymer Degradation and Stability* 94 (1), 1-6.
70. Hong, H.J. et al. (2008) Accelerated wound healing by smad3 antisense oligonucleotides-impregnated chitosan/alginate polyelectrolyte complex. *Biomaterials* 29 (36), 4831-4837.
71. Rao, K. et al. (2008) Synthesis and characterization of poly(acrylamidoglycolic acid) grafted onto chitosan and its polyelectrolyte complexes with hydroxyapatite. *Reactive and Functional Polymers* 68 (5), 943-953.
72. De Oliveira, H.C.L. et al. (2008) Chitosan-poly(acrylic acid) polyelectrolyte complex membranes: preparation, characterization and permeability studies. *Journal of Biomaterials Science-Polymer Edition* 19 (2), 143-160.
73. Bigucci, F. et al. (2008) Chitosan/pectin polyelectrolyte complexes: Selection of suitable preparative conditions for colon-specific delivery of vancomycin. *European Journal of Pharmaceutical Sciences* 35 (5), 435-441.
74. Kim, S.J. et al. (2004) Synthesis and characteristics of polyelectrolyte complexes composed of chitosan and hyaluronic acid. *Journal of Applied Polymer Science* 91 (5), 2908-2913.
75. Fukuda, H. and Kikuchi, Y. (1978) In vitro clot formation on polyelectrolyte complexes of sodium dextran sulfate with chitosan. *Journal of Biomedical Materials Research* 12 (4), 531-539.
76. Ii'ina, A.V. and Varlamov, V.P. (2005) Chitosan-based polyelectrolyte complexes: A review. *Applied Biochemistry and Microbiology* 41 (1), 5-11.

77. Mao, H.Q. et al. (2001) Chitosan-DNA nanoparticles as gene carriers: synthesis, characterization and transfection efficiency. *Journal of Controlled Release* 70 (3), 399-421.

78. Lopez-Leon, T. et al. (2005) Physicochemical characterization of chitosan nanoparticles: electrokinetic and stability behavior. *Journal of Colloid and Interface Science* 283 (2), 344-351.

79. Huang, Y. et al. (2005) In vitro characterization of chitosan-gelatin scaffolds for tissue engineering. *Biomaterials* 26 (36), 7616-7627.

80. Xia, W.Y. et al. (2004) Tissue engineering of cartilage with the use of chitosangelatin complex scaffolds. *Journal of Biomedical Materials Research Part BApplied Biomaterials* 71B (2), 373-380.

81. Hsieh, C.Y. et al. (2005) Preparation of gamma-PGA/chitosan composite tissue engineering matrices. *Biomaterials* 26 (28), 5617-5623.

82. Peng, L. et al. (2006) Preparation and evaluation of porous chitosan/collagen scaffolds for periodontal tissue engineering. *Journal of Bioactive and Compatible Polymers* 21 (3), 207-220.

83. Li, Z.S. and Zhang, M.Q. (2005) Chitosan-alginate as scaffolding material for cartilage tissue engineering. *Journal of Biomedical Materials Research Part A* 75A (2), 485-493.

84. Kim, T.H. et al. (2007) Chemical modification of chitosan as a gene carrier in vitro and in vivo. *Progress in Polymer Science* 32 (7), 726-753.

85. Fukuda, H. and Kikuchi, Y. (1977) Polyelectrolyte complexes of sodium dextran sulfate with chitosan, .2. *Makromolekulare Chemie-Macromolecular Chemistry and Physics* 178 (10), 2895-2899.

86. Kabanov, V.A. and Zezin, A.B. (1984) Soluble interpolymeric complexes as a new class of synthetic poly-electrolytes. *Pure and Applied Chemistry* 56 (3), 343-354.

87. Chen, Y. et al. (2003) Chitosan-dextran sulfate nanoparticles for delivery of an anti-angiogenesis peptide. *Letters in Peptide Science* 10 (5-6), 621-629.

88. Chen, Y. et al. (2008) Development of a chitosan-based nanoparticle formulation for delivery of a hydrophilic hexapeptide, dalargin. *Biopolymers* 90 (5), 663-670.

89. Huang, M. and Berkland, C. (2009) Controlled release of repifermin (R) from polyelectrolyte complexes stimulates endothelial cell proliferation. *Journal of Pharmaceutical Sciences* 98 (1), 268-280.

90. Drogoz, A. et al. (2008) Towards biocompatible vaccine delivery systems: Interactions of colloidal PECs based on polysaccharides with HIV-1 p24 antigen. *Biomacromolecules* 9 (2), 583-591.

91. Sarmento, B. et al. (2007) Oral bioavailability of insulin contained in polysaccharide nanoparticles. *Biomacromolecules* 8 (10), 3054-3060.

92. Tiyaboonchai, W. and Limpeanchob, N. (2007) Formulation and characterization of amphotericin B-chitosan-dextran sulfate nanoparticles. *International Journal of Pharmaceutics* 329 (1-2), 142-149.

93. O'Neil, M.J. (2001) *The Merck index : an encyclopedia of chemicals, drugs, and biologicals*, Merck.

94. Crisp, G.T. and Jiang, Y.L. (2001) Intramolecular hydrogen bonding of (+)-biotin and biotin derivatives in organic solvents. *Arkivoc* 2.

95. Musashi, Y. et al. (1995) Structure and bonding nature of carboxyimidazolidone, a model of carboxybiotin - Ab-Initio Mo/Mp4, Sd-Ci, and Ccd studies. *Journal of the American Chemical Society* 117 (45), 11320-11326.

96. Wood, H.G. and Kumar, G.K. (1985) Transcarboxylase - its quaternary structure and the role of the biotinyl subunit in the assembly of the enzyme and in catalysis. *Annals of the New York Academy of Sciences* 447 (JUN), 1-22.

97. Savage, D. et al. (1992) *Avidin-biotin chemistry: A handbook*, Pierce Chemical Co.

98. Lim, S.C. et al. (2009) Fabrication and characterization of an OTFT-based biosensor using a biotinylated F8T2 polymer. *ETRI Journal* 31 (6), 647-652.

99. Diamandis, E.P. and Christopoulos, T.K. (1991) The biotin (strept) avidin system - principles and applications in biotechnology. *Clinical Chemistry* 37 (5), 625-636.

100. Lesch, H.P. et al. (2010) Avidin-biotin technology in targeted therapy. *Expert Opinion on Drug Delivery* 7 (5), 551-564.

101. Clapper, J.D. et al. (2008) Biotinylated biodegradable nanotemplated hydrogel networks for cell interactive applications. *Biomacromolecules* 9 (4), 1188-1194.

102. Meng, Q.B. et al. (2007) Aggregation of biotinylated polymeric microspheres induced by interaction with avidin. *Pure and Applied Chemistry* 79 (9), 1575- 1582.

103. Xiong, M.P. et al. (2007) Biotin-triggered release of poly(ethylene glycol)-avidin from biotinylated polyethylenimine enhances in vitro gene expression. *Bioconjugate Chemistry* 18 (3), 746-753.

104. Ben-Shabat, S. et al. (2006) PEG-PLA block copolymer as potential drug carrier: Preparation and characterization. *Macromolecular Bioscience* 6 (12), 1019-1025.

105. Salem, A.K. et al. (2001) Synthesis and characterisation of a degradable poly(lactic acid)-poly(ethylene glycol) copolymer with biotinylated end groups. *Biomacromolecules* 2 (2), 575-580.

106. Shi, X.W. et al. (2008) Chitosan biotinylation and electrodeposition for selective protein assembly. *Macromolecular Bioscience* 8 (5), 451-457.

2

CHAPTER

SYNTHESIS AND OPTIMIZATION OF CHITOSAN NANOPARTICLES: POTENTIAL APPLICATIONS IN NANOMEDICINE AND BIOMEDICAL ENGINEERING

Arezou Ghadi[1], Soleiman Mahjoub[2,3], Fatemeh Tabandeh[4], Farid Talebnia[1]

[1]Faculty of Chemical Engineering, Noshirvani University of Technology, Babol, Iran.

[2]Fatemeh Zahra Infertility and Reproductive Health Research Center, Babol University of Medical Sciences, Babol, Iran.

[3] Department of Biochemistry & Biophysics, Faculty of Medicine, Babol University of Medical Sciences, Babol, Iran.

[4]National Institute of Genetic Engineering and Biotechnology (NIGEB), Tehran, Iran.

ABSTRACT

Background: Chitosan nanoparticles have become of great interest for nanomedicine, biomedical engineering and development of new therapeutic drug release systems with improved bioavailability, increased specificity and sensitivity, and reduced pharmacological toxicity. The aim of the present study was to synthesis and optimizes of the chitosan nanoparticles for industrial and biomedical applications.

Methods: Fe3O4 was synthesized and optimized as magnetic core nanoparticles and then chitosan covered this magnetic core. The size and morphology of the nano-magnetic chitosan was analyzed by scanning electron microscope (SEM). Topography and size distribution of the nanoparticles were shown with two-dimensional and three-dimensional images of atomic force microscopy (AFM). The nanoparticles were analyzed using transmission electron microscopy (TEM).

Results: The chitosan nanoparticles prepared in the experiment exhibited white powder shape. The SEM micrographs of the nano-magnetic chitosan showed that they were approximately uniform spheres. The unmodified chitosan nanoparticles composed of clusters of nanoparticles with sizes ranging from 10 nm to 80 nm. AFM provides a three-dimensional surface profile. The TEM image showed physical aggregation of the chitosan nanoparticles.

Conclusion: The results show that a novel chitosan nanoparticle was successfully synthesized and characterized. It seems that this nanoparticle like the other chitosan nano particles has potential applications for nanomedicine, biomedical engineering, industrial and pharmaceutical fields.

INTRODUCTION

In recent years, the identification of the unique properties of nanoparticles has allowed their application in many fields such as biomedicine, drug and gene delivery, tumor detection, tumor vaccines, tissue engineering, MRI contrast enhancement, sensor development, fluorescent biological labels, detection of pathogens, detection of protein, separation and purification of cells and biological molecules, probing of DNA structure, environmental remediation and water purification (1-7). They are used in the diagnosis and therapeutics due to their unique properties of having small size, large surface area to volume ratio, stability over high temperatures, translocation into the cells and high reactivity to the living cells. They are available in different sizes and shapes due to their ability to react and agglomerate with other nanoparticles in their surroundings.

They also exhibit exceptional optical properties making them capable of producing quantum effects suitable for imaging applications (1-3). Most commonly studied metallic nanoparticles include gold, silver, aluminum, zinc, iron and titanium oxide nanoparticles (8). However, due to their “nano” size, their entry is easily facilitated into different cells posing one of the greatest problems in using these nanoparticles for targeted delivery to specific tissues. To resolve this problem, researchers have been conjugating these nanoparticles with various biomolecules and ligands to develop strategies for targeted delivery. The large surface-area-to-volume ratio of a nanoparticle allows it to serve as an efficient carrier of biomolecules. This feature has resulted in the development of many biomolecule-nanoparticle (bio-NP) hybrids for biomedical applications in the diagnosis and localized treatment of diseases. The benefits of magnetic nano-carriers are resistant against chemical and microbial environment, high mechanical resistance and good thermal stability (8-10). Chitosan [(1, 4)-2-amino-2-deoxy-D-glucan] is a linear polyaminosaccharide obtained by N-deacetylation of chitin, the second most abundant natural bio-polymer, after cellulose. It can be easily processed in diverse forms, such as films, threads tablets, membranes and microparticles/ nanoparticles, allowing the design of a variety of medical and pharmacological devices adaptable to end purposes. In particular, in medicine, chitosan may be useful in bandages to reduce bleeding and as an antibacterial agent can also be used to help deliver drugs through the skin. It is also used in the development of chitosan drug control releasing systems including chitosan sponges, chitosan film, chitosan beads, chitosan microbeads (microspheres) and chitosan nanoparticles. In other words, nano-magnetic is the core and chitosan polymer covering the magnetic cores. The process is called coating. These magnetic nanoparticles can be used as powerful carriers for enzyme immobilization (11, 12). Chitosan nanoparticles have an antitumor role through improving the body's immune function (13-18). Despite the well known advantages of exploiting chitosan in these fields, an additional work needs to be done to optimize chitosan's formulations and enhance its physicochemical properties for different uses. The aim of the present study was to synthesis and optimize the formulation parameters of chitosan magnetic nanoparticles for biomedical engineering and pharmaceutical applications and present a method for synthesizing of novel chitosan magnetic nanoparticles.

METHODS

This experimental study was done in the Biochemistry Department of Babol University of Medical Sciences, Babol and National Institute of Genetic Engineering and Biotechnology, Tehran, Iran.

Synthesis and Optimization of Magnetic Core Nanoparticles

In this study, Fe_3O_4 was synthesized and optimized as magnetic core nanoparticles. For the preparation of magnetic core nanoparticles, aqueous solutions of Fe (II) /Fe (III) chloride (molar ratio 1:2) was prepared and added to an oxygen free container while N_2 was flown in it. The solution was stirred for 30 sec and then, 20 mL NH3 was quickly added to the solution and pH was adjusted at 9. After 45 min, the prepared nanoparticles were stuck to the magnet and separated easily. The synthesized magnet nanoparticles were washed with deionized water at pH7, and dried at 37°C. The amount of 0.02 g magnet nanoparticles were dissolved in 50 mL deionized water and combined with 50 mL tri-sodium citrate solution. The result solution was sonicated for 30 min. This procedure was confirmed after optimization of the method.

Synthesis and Optimization of Chitosan Nanoparticles

The amount of 500 mg of chitosan (medium molecular weight and 85% deacetylated, Sigma Chemical, St. Louis, USA) was dissolved in 50 ml of 1% acetic acid solution and stirred at 1000 rpm for 25 min at room temperature until the solution became clear. The resulting solution was sonicated and then titrated by addition of NaOH or HCL solution adjusted to pH5 and filtered using 0.2μ mesh. For coating process, 5 ml nano-magnetic solution was added to 75 mL deionized water and sonicated for 10 min. Then, chitosan solution was added and sonicated for 5 min. The resulting solution was clear.

Electron Microscopy Analysis

After the preparation of the synthesized chitosan nanoparticles, the characterization of the nanoparticle was examined by scanning electron microscope (SEM) using Stereoscan (model S360 brand SEM – Leica Cambridge, Cambridge, UK). Also, the size and morphology of the nano-magnetic chitosan was observed by transmission electron microscopy (TEM) (Philips CM200 EFG, FEI Company, Eindhoven, Netherlands). In addition, the pictures of atomic force microscopy (AFM) were taken using an Autoprobe CP Research AFM system (model AP- 2001, Thermomicroscopes, USA).

RESULTS

Chitosan nanoparticles prepared in the experiment exhibited white powder shape. The morphology of chitosan nanoparticles was analyzed by scanning electron microscope (SEM). The unmodified chitosan nanoparticles composed of clusters of nanoparticles (figure 1).

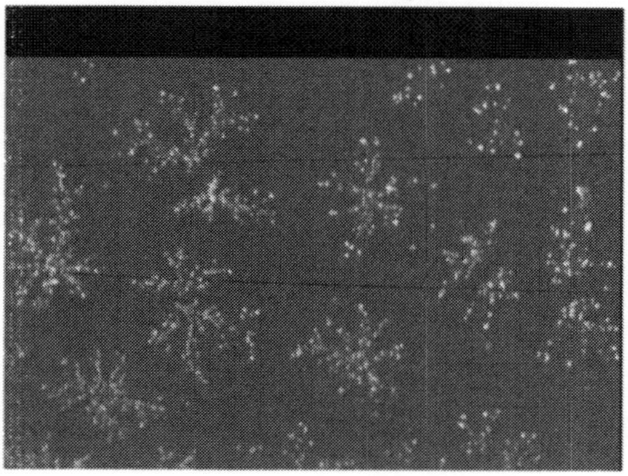

Figure 1 Scanning electron microsopy (SEM) micrograph of the chitosan nanoparticles.

The scanning electron micrographs of the chitosan magnetic nanoparticles showed that they were approximately uniform spheres. Unlike electron microscopes, which provide a two-dimensional projection or a two-dimensional image of a sample, AFM provides a three-dimensional surface profile. Although the lateral dimensions are influenced by the shape of the probe, the height measurements can provide the height of nanoparticles with a high degree of accuracy and precision. Figure 2 showed that the analysis of size distribution of the chitosan nanoparticles by atomic force microscopy. As shown in the histogram of AFM (Fig 2-C), the size distribution of the nano-magnetic chitosan particles is between 10-80 nm. The observation from transmission electron microscope (TEM) gave us information on the particle shape and the determination of particle size. Typical TEM micrograph of the chitosan nanoparticles was shown in figure 3. In the present study, TEM images showed physical aggregation of the chitosan nanoparticles. According to electron microscope, the analysis of the chitosan magnetic nanoparticles diameter of the essentially monodispersed Fe3O4 particles ranged between 5 am to 8 nm, and the diameter of the nano-magnetic chitosan ranged between 10 nm to 80 nm. The Fe3O4 particles were well-coated by chitosan.

DISCUSSION

In the present study, we successfully synthesized and optimized a novel magnetic chitosan nanoparticle as potential applications for medical, pharmaceutical and industrial fields. The applications of magnetic nanoparticles in medical and pharmaceutical fields can be rationalized as follows: first, the magnetic chitosan nanoparticles are a class of intrinsically ordered magnetic materials. As their magnetic signal is generated by the application of external strong magnetic field, it far exceeds that signal from any of the known molecules. This makes them readily identified by a magnetic sensor or magnet from other molecules.

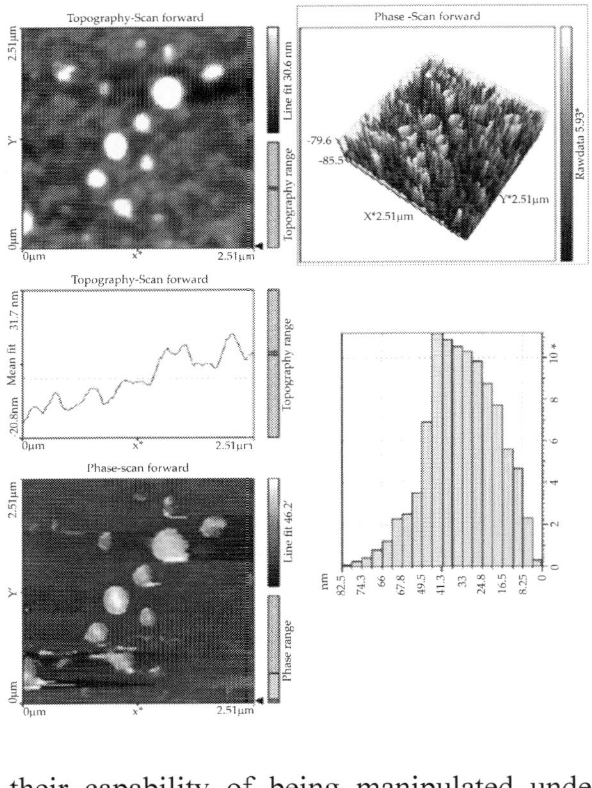

Second, their capability of being manipulated under an external magnetic field provides controllable means of magnetically tagging biomolecules, leading to potentially highly sensitive biosensing highly efficient bioseparation, and magnetic resonance imaging (MRI) contrast enhancement, as well as site-specific drug delivery (5, 19, 20).

Chitosan nanoparticles have become of great interest as polymeric platforms for the development of new pharmacological and therapeutic drug release systems with improved biodistribution and increased specificity and sensitivity, and reduced pharmacological toxicity. Chitosan nanoparticles have been found appropriate for non-invasive routes of drug administration: nasal, oral, ocular and pulmonary routes. These applications are facilitated by the absorption-enhancing effect of chitosan. Additionally, chitosan nanoparticles have been proposed as non-viral vectors in gene therapy and have shown adjuvant effect in vaccines. Absorption and bioavailability of drug encapsulated into chitosan nanoparticles can be improved, so they can be used to deliver gene drugs, protein drugs and other compounds and can protect them effectively from enzyme degradation in vivo (21, 22).

As a carrier, nanochitosan can improve DNA and drug bioavailability, resistance in vivo to enzyme drop solutions, enhance controlled sustained release of biomaterials, reduce toxicity, and be prepared under mild conditions without the use of an organic solvent, thereby avoiding DNA and drug destruction, as well as preventing residual solvent remaining after the preparation process (23).

Chitosan nanoparticles can be used to alter protein loading and adjust the value of each parameter during preparation. They also have high stability, high protein packing efficiency, can be prepared as a lyophilized powder, and are easy to store and transport (24). In our study, we prepared the magnetic chitosan nanoparticles as white powder and confirmed characteristics of the nano product using scanning electron microscope, transmission electron microscopy, and atomic force microscopy. The nanoparticles were stored as lyophilized powder for various applications. Most biological macromolecules, including proteins, cannot enter cells because of the selective permeability of the cell membrane, which hampered application of tumor vaccines (25).

Conventional colloidal carriers are rapidly removed from the bloodstream by the reticuloendothelial system (RES), which is a part of the mononuclear system (MPS) after intravenous administration. Nanoparticulate systems have been used to improve the blood circulation time and tumor targeting efficacy, because the tumor vascular permeability allows the penetration of particles up to 400 nm in diameter. According to our study, the unmodified chitosan nanoparticles composed of clusters of nanoparticles with sizes ranging from 10 nm to 80 nm. The antitumor efficacy of chitosan nanoparticles administered by intravenous injection is probably attributed to their little particle size. Particle size was proven as an important feature related to obtaining optimal in vitro efficacy of chitosan nanoparticles. Particle size also had a crucial impact on the in vivo fate of a drug delivery system. Decreasing particle size could increase the surface-to volume ratio and specific surface area, which could increase the dissolution and thus increase the bioavailability of poorly water soluble molecules. The smaller size particles seem to have efficient interfacial interaction with the cell membrane compared to larger size particles due to the endocytosis of small size particles. Small size particles could improve the efficacy of the particle-based oral drug delivery systems. Also, the use of particle size reduction to increase the oral bioavailability of drugs has been obtained. Nanoparticles can prolong the blood half-life of drugs and increase the efficacy by intravenous injection (26, 27).

The major disadvantage of most drugs for tumor chemotherapy is their relative nonspecificity. The drugs are administered intravenously for general system distribution, resulting in deleterious side effects as they attack normal, healthy cells in addition to the target tumor cells. Preferably, the drugs should be localized to the tumor site. Chitosan nanoparticles are now being modified for sustained/ controlled release and targeting. The attachment of drugs to magnetic particles can be used to reduce drug dose and potential side effects on healthy tissue and the costsassociated with drug treatment. The size, charge and surface chemistry of the magnetic particles are particularly important in affecting both blood circulation time as well as the bioavailability of the particles within the body (4, 13, 18, 19). In the absence of any surface coating, nano-magnetic particles have hydrophobic surfaces with a large surface area to volume ratio. Due to hydrophobic interactions between the particles, they tend to agglomerate forming large clusters. In the present study, the TEM image showed the physical aggregation of the nanomagnetic chitosan. There are two possible reasons for this behavior. First, as the Fe_3O_4 nanoparticles were present in the pores of the chitosan nanoparticles, this may be cross-link between chitosan chains which caused the aggregation of magnetic nanoparticles. Second, the diameter of the particles was too small, and the surface energy was very high, which may cause the physical aggregation. Although some observed aggregation could be caused by drying during TEM sample preparation.

In our study, Fe_3O_4 was synthesized and optimized as magnetic core nanoparticles and then chitosan covered this magnetic core as a shell. We showed the high magnetic characteristics of the synthesized nano chitosan. Magnetic nanoparticles respond resonantly to an alternating magnetic field, allowing the transfer of magnetic energy to the particles as a form of heat. This has been proposed to be one of the key approaches to successful cancer therapy in the future (28). Also, nanomaterials combined with tumor antigens allow tumor vaccines to have stronger biocompatibility, permeability, and targeting properties (1). In the future, nanotechnology makes more progress, and more new nanomaterials continue to appear, they will be used increasingly in the fields of tumor vaccines, gene and drug carriers. Finally as a conclusion, we successfully synthesized and characterized a novel chitosan nanoparticle using the optimization of condition. It is believed that the synthesized novel chitosan nanoparticles can be considered for different biomedical applications.

REFERENCES

1. Li XD, Gao JY, Yang Y, et al. Nanomaterials in the application of tumor vaccines: advantages and disadvantages. OncoTargets Ther 2013; 6: 629-34.
2. Saini R, Saini S, Sharma S. Nanotechnology: The Future Medicine. J Cutan Aesthet Surg 2010; 3: 32-3.
3. Salata O. Applications of nanoparticles in biology and medicine. J Nanobiotech nology 2004; 2: 3.
4. Wang JJ, Zeng ZW, Xiao RZ, et al. Recent advances of chitosan nanoparticles as drug carriers. Int J Nanomedicine 2011; 6: 765-74.
5. Gupta AK, Gupta M. Synthesis and surface engineering of iron oxide nanoparticles for biomedical applications. Biomaterials 2005; 26: 3995-4021.
6. Tratnyek PG, Johnson RL. Nanotechnologies for environmental

cleanup. Nano today 2006; 1: 44-8.

7. Savage N, Diallo MS. Nanomaterials and water purification: opportunities and challenges. J Nanopart Res 2005; 7: 331-42.
8. Katz E, Willner I. Integrated nanoparticle-biomolecule hybrid systems: synthesis, properties, and applications. Angew Chem Int Ed Engl 2004; 43: 6042-108.
9. Tiwari PM, Vig K, Dennis VA, Singh SR. Functionalized gold nanoparticles and their biomedical applications. Nanomaterials 2011; 1: 31-63.
10. Ito A, Shinkai M, Honda H, Kobayashi T. Medical application of functionalized magnetic nanoparticles. J Biosci Bioeng 2005; 100: 1-11.
11. Kim JY, Lee JK, Lee TS, Park WH. Synthesis of chitooligosaccharide derivative with quaternary ammonium group and its antimicrobial activity against Streptococcus mutans. Int J biol macromol 2003; 32: 23-7.
12. Orrego CE, Salgado N, Valencia JS, et al. Novel chitosan membranes as support for lipases immobilization: Characterization aspects. Carbohydrate Polymers 2010; 79: 9-16.
13. 13.Aruna U, Rajalakshmi R, Indira Muzib Y, et al. Role of Chitosan Nanoparticles in Cancer Therapy. Int J Innovative Pharmaceutical Res 2013; 4: 318-24.
14. Cao J, Zhou NJ. Progress in antitumor studies of chitosan. Chin J Biochem Pharm 2005; 26(2): 127.
15. Zhou SH, Hong Y, Fang GJ. Preparation, characterization and anticancer effect of chitosan nanoparticles. J Clin Rehab Tiss Engin Res 2007; 11: 9688-91.
16. Torzsas T, Kendall C, Sugano M, Iwamoto Y, Rao A. The influence of high and low molecular weight chitosan on colonic cell proliferation and aberrant crypt foci development in CF1 mice. Food Chem Toxicol 1996; 34: 73-7.
17. Maeda Y, Kimura Y. Antitumor effects of various low-molecular-weight chitosans are due to increased natural killer activity of intestinal intraepithelial lymphocytes in sarcoma 180-bearing mice. J Nutr 2004; 134: 945-50.

18. Fang GJ, Hong Y, Jiang YY. Comparison of antitumor effects of chitosan nanoparticles from different sources in vitro. J Clin Rehab Tiss Engin Res 2007; 11: 9696-9.
19. Dobson J. Magnetic nanoparticles for drug delivery. Drug Develop Res 2006; 67: 55-60.
20. Neuberger T, Schopf B, Hofmann H, Hofmann M, Rechenberg B. Superparamagnetic nanoparticles for biomedical applications: Possibilities and limitations of a new drug delivery system. J Magn Magn Mater 2005; 293: 483-96.
21. Peniche H, Peniche C. Chitosan nanoparticles: a contribution to nanomedicine. Polymer Int 2011; 60: 883-9.
22. Shi SF, Jia JF, Guo XK, et al. Biocompatibility of chitosan-coated iron oxide nanoparticles with osteoblast cells. Int J Nanomedicine 2012; 7: 5593-602.
23. Chen MC, Mi FL, Liao ZX, Hsiao CW, Sonaje K, Chung MF, Hsu LW, Sung HW. Recent advances in chitosan-based nanoparticles for oral delivery of macromolecules. Adv Drug Deliv Rev 2013; 65: 865-79.
24. Chen F, Shi Z, Neoh KG, Kang ET. Antioxidant and antibacterial activities of eugenol and carvacrol-grafted chitosan nanoparticles. Biotechnol Bioeng 2009; 104: 30-9.
25. Wadhwa PD, Zielske SP, Roth JC. Cancer gene therapy: scientific basis. Annu Rev Med 2002; 53: 437-52.
26. Qi L, Xu Z. In vivo antitumor activity of chitosan nanoparticles. Bioorg Med Chem Lett 2006; 16: 4243-5.
27. Wadhwa PD, Zielske SP, Roth JC, et al. Cancer gene therapy: scientific basis. Annu Rev Med 2002; 53: 437-52.
28. Yigit MV, Moore A, Medarova Z. Magnetic nanoparticles for cancer diagnosis and therapy. Pharm Res 2012; 29: 1180-8.

3

CHAPTER

CHITOSAN NANOPARTICLES SYNTHESIS CAUGHT IN ACTION USING MICRODROPLET REACTIONS

Vivek Kamat, Dhananjay Bodas & Kishore Paknikar

Nanobioscience, Agharkar Research Institute, GG Agarkar Road, Pune 411 004 India

ABSTRACT

The ionic gelation process for the synthesis of chitosan nanoparticles was carried out in microdroplet reactions. The synthesis could be stopped instantaneously at different time points by fast dilution of the reaction mixture with DI water. Using this simple technique, the effect of temperature and reactant concentrations on the size and distribution of the nanoparticles formed, as a function of time, could be investigated by DLS and SEM. Results obtained indicated very early (1–5 s) nucleation of the particles followed by growth. The concentration of reactants, reaction temperature as well as time, were found to (severally and collectively) determine the size of nanoparticles and their distribution. Nanoparticles obtained at 4 °C were smaller (60–80 nm) with narrower size distribution. Simulation experiments using Comsol software showed that at 4 °C 'droplet synthesis' of nanoparticles gets miniaturised to 'dropletcore synthesis', which is being reported for the first time.

INTRODUCTION

Chitosan nanoparticles are biocompatible, relatively non-toxic, biodegradable, and cationic in nature1, 2. Thus, they are well accepted in biomedical applications such as drug delivery3–5. Ionic gelation is the most commonly used method for synthesising chitosan nanoparticles6. In this method, chitosan precursors are cross-linked using sodium tripolyphosphate (TPP). The method typically yields large sized (100–300 nm) particles with a high degree of polydispersity. Even though ionic gelation is a widely used method and factors governing the size and dispersivity of chitosan nanoparticles (such as the concentration of reactants, temperature, pH, and the level of deacetylation) are well known7 our basic understanding of the process at mechanistic level is poor. In the ionic gelation process, TPP crosslinks randomly oriented chitosan molecules, which, in turn, are connected to other similarly cross-linked moieties. Such intra- and inter- molecular cross-linking is rather uncontrolled and leads to polydispersity in the synthesised preparation. Here, we hypothesised that using confined reaction volumes of microdroplets and at preset temperatures; it should be possible to exert good control over the kinetics of nanoparticles synthesis if the reactions could be arrested rapidly at desired time points. Such an approach could enable time-lapsed capture of the synthesis process and provide us an opportunity to understand it at a more fundamental level.

Thus, in our microdroplet experiments, synthesis reactions were arrested at varying time points to monitor the nucleation and growth of developing nanoparticles using DLS and SEM. For this purpose a simple yet novel technique, viz. fast dilution of the reaction mixture with DI water was employed. In a separate experiment, we had determined that with a 500-fold dilution, the reaction pH increased sharply from 2.84 ± 0.04 to 5.50 ± 0.05. Such a rise in pH could exert a strong endothermic effect8–10, resulting in extensive (90%) de-protonation of chitosan11. Dilution could also result in increased intra-molecular distances, reducing the possibility of molecular interaction. The synthesis of chitosan nanoparticles tends to be arrested by DI water dilution owing to these reasons.

RESULTS AND DISCUSSION

It is evident from the data presented in Fig. 1a–c, that with an increase in the concentration of chitosan from 0.2 to 0.8 mg/mL a concomitant increase in the number of nanoparticles formed could be seen. The number of the hyphen should appear after nano (nano-particles) also increased with rising reaction temperature and time. Nanoparticles of smallest size (71 nm) were obtained when the reaction was carried out with 0.4 mg/mL chitosan at 4 °C. The number of nanoparticles was maximum (5.16E8) with 0.8 mg/mL chitosan (at 35 °C). The narrowest size distribution of the particles was seen when the chitosan concentration was 0.4 mg/mL and the reaction was performed at 4 °C (Fig. 3c). It is clear from these data that an interplay of factors like temperature, time and concentration of reactants determines the size, distribution and the number of particles formed.

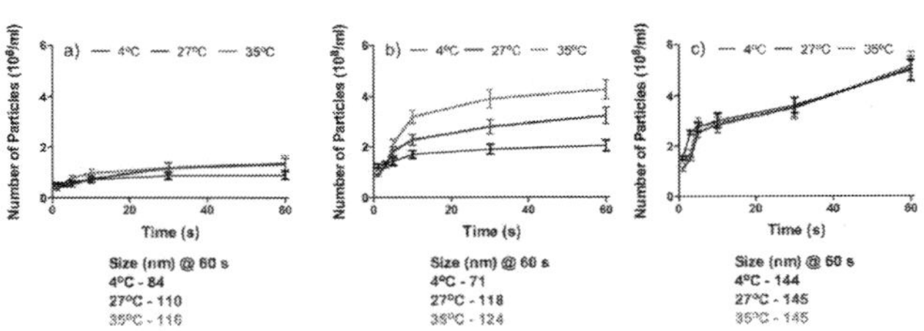

Figure 1: Time- and temperature- dependent change in the number of nanoparticles synthesized using different chitosan concentrations (mg/mL), viz. (a) 0.2; (b) 0.4 and (c) 0.8.

Interestingly, the number of nanoparticles formed in the initial few seconds (1–5 s) was higher at 4 °C than at 27 °C and 35 °C, irrespective of the chitosan concentration used. It was also seen that the magnitude of difference in the number of nanoparticles formed was highest when the chitosan concentration used was 0.4 mg/mL. Therefore, in further studies chitosan concentration was maintained at 0.4 mg/mL while the reaction temperatures were varied. The time-lapsed capture of the nanoparticles synthesis process was achieved by arresting the reactions at different time points and imaging the preparation by scanning electron microscopy (SEM). SEM images presented in Fig. 2a show that at 4 °C, after 1 s, the chitosan polymer was seen in the crystalline state, which rapidly formed a large number of small and monodispersed particles at

3 s. This observation indicated that the nucleation phase was initiated between 1–3 s, followed by growth phase. The image after 10 s (Fig. 2a) shows fully-grown chitosan nanoparticles. At 27 °C (Fig. 2b), unreacted chitosan was seen in 1 s, along with very few nanoparticles. At 3 s, nucleation of the nanoparticles was initiated on the surface of chitosan polymer (seen as bright spots decorating the polymer). At 5 s, the conversion of the polymer to well defined nascent nanoparticles was visible which subsequently underwent growth. In the reactions carried out at 35 °C (Fig. 2c), conversion of chitosan polymer to nanoparticles was instantaneous. Simultaneous and seamless nucleation and growth followed thereafter. Such apparently uncontrolled growth of nanoparticles formed agglomerates at 35 °C resulting in a larger size (130–190 nm) and broader size distribution, as seen after 10 s. On the other hand, nanoparticles obtained at 4 °C were smaller (60–80 nm) and with narrower size distribution.

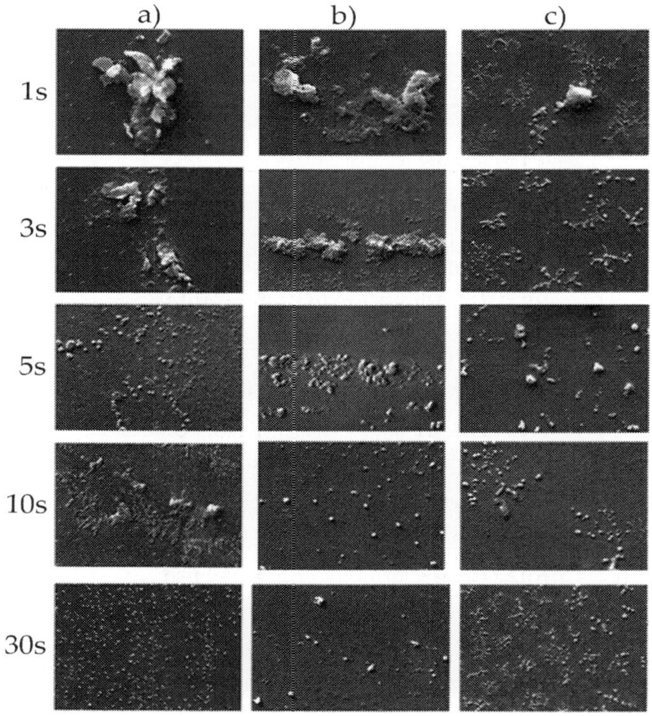

Figure 2: SEM of chitosan nanoparticles captured at varying time points and temperatures.

To discern the mechanism of chitosan nanoparticles synthesis at 4 °C, simulation studies were carried out using Comsol Multiphysics software. The studies focused on understanding the heat and solute transport across the droplet. Figure 3a shows a simulation model of a (chitosan + TPP) droplet diluted with DI water. The properties of chitosan + TPP were allocated to the droplet whereas DI water properties were ascribed to the surrounding medium.

To study heat transfer, the temperature values assigned to the droplet, and the surrounding medium were 4 °C and 30 °C, respectively. Assigned reactant concentrations of the droplet and the medium were 0.4 mg/mL and 0 mg/mL, respectively. Heat and solute transfers were simulated over 0 to 5 s (the experimentally observed period crucial in the formation of chitosan nanoparticles) and were monitored at the core and periphery of the droplet.

The results of simulation studies are shown in Fig. 3b. Chitosan droplet (at 4 °C) upon dilution with DI water (at 30 °C), experiences a sharp rise in the temperature of the droplet core as well as the periphery in 100 ms. At this stage, the chitosan concentration at the droplet core was 0.4 mg/mL, which gradually decreased to 0.25 mg/mL in 2000 ms. In contradistinction, the chitosan concentration in the droplet periphery dropped instantaneously to 0.1 mg/mL, and lowered further to 0.05 mg/mL in 2000 ms.

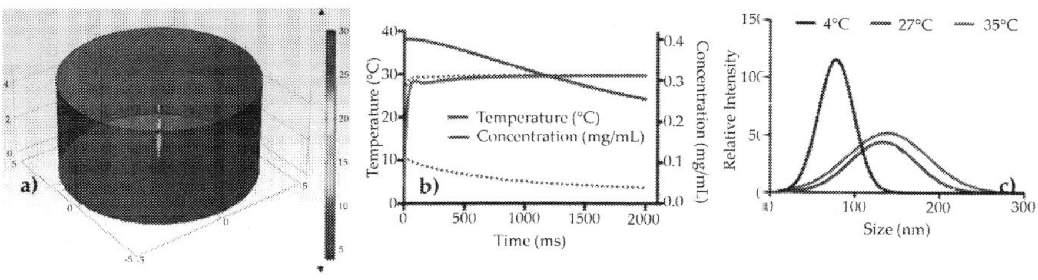

Figure 3: (a) Simulation model of the droplet and the surrounding medium. (b) Simulation profile showing the variation of temperature and reactant concentration as a function of time. Dotted and continuous line represent temperature and concentration at the droplet periphery and the core respectively; (c) Effect of temperature on the size distribution of chitosan nanoparticles (chitosan concentration used 0.4 mg/mL).

The ultrafast dilution of the reactants (owing to diffusion) in the droplet periphery as observed here precludes any possibility of fresh nucleation and/or growth. On the other hand, steep rise the in the droplet core temperature along with low molecular mobility favour faster reaction kinetics. A gradual decrease in chitosan concentration in the core over a period of 2000 ms suggests well-controlled nucleation and growth, leading to the smaller size of nanoparticles formed with a narrower distribution. In short, at 4 °C the droplet synthesis translates into further miniaturised and refined droplet-core synthesis. The experimental validation of these simulation results showed that smaller size and narrower size distribution is indeed obtained at 4 °C as compared to 27 °C and 35 °C (Fig. 3c). This study provides a critical insight into ionic gelation process of chitosan nanoparticles formation, caught in action, for the first time. It is evident from the data presented that the chitosan nanoparticles nucleate very early (1–5 s) and quickly enter the growth phase. The study shows that the concentration of reactants, temperature, and time, severally and collectively determine the fate of the synthesis process in terms of the size and distribution of particles formed. For example, obtaining smaller sized nanoparticles could be at the expense of distribution and/ or number of particles. Our simulation experiments point to a new concept, i.e. the quality of nanoparticles relies on the temperature dependent mechanistic pathway rather than the temporal endpoint. Finally, these studies indicate that when it comes to obtaining superior quality polymeric nanoparticles, the secret lies at the core of the droplet!

METHODS

Chitosan (CS) from crab shell (molecular weight 150000 kD, degree of deacetylation 75%) was procured from Sigma-Aldrich, USA. TPP and glacial acetic acid were purchased from Loba Chemie, India and Qualigens, India, respectively. Freshly prepared, deionized (DI) water (pH 6.12 ± 0.03) was used as a diluent. Different sets of reactions were performed where chitosan and TPP concentrations were altered in the range of 0.2–0.8 mg/mL and 0.08–0.32 mg/mL, respectively keeping chitosan:TPP ratio constant (2.5:1). The reaction temperatures used were 4 °C, 27 °C and 35 °C. Briefly, two (1 μL) droplets of chitosan and TPP were sequentially aspirated using an electronic pipette (Thermo Scientific Novus Finnpipette), allowed to react in the pipette tip and the reactions stopped at varying times (1, 3, 5, 10, 30 and 60 s) by 500-fold dilution with DI water. The number of nanoparticles formed and their sizes were determined by dynamic light scattering (DLS) method using Nanosight LM10 system (Malvern, UK).

The nanoparticles formed at different times were also visualized by scanning electron microscopy (Zeiss EVO MA15, UK).

REFERENCES

1. Wang, J. *et al.* Recent advances of chitosan nanoparticles as drug carriers. *Int. J. Nanomed.* **6,** 765–774 (2006).
2. Keawchaoon, L. & Yoksan, R. Preparation, characterisation and in-vitro release study of carvacrol-loaded chitosan nanoparticles. *Colloids and Surfaces B: Biointerfaces* **84,** 163–171 (2011).
3. Patel, T., Zhou, J. J. P. & Saltzman, W. Polymeric nanoparticles for drug delivery to the central nervous system. *Advanced Drug Delivery Reviews* **64,** 701–705 (2012).
4. Prabaharan, M. Chitosan-based nanoparticles for tumor-targeted drug delivery. *Intl. J. of Biological Macromolecules* **72,** 1313–1322 (2015).
5. Agnihotri, S., Mallikarjuna, N. & Aminabhavi, T. Recent advances on chitosan-based micro- and nanoparticles in drug delivery. *J. Controlled Release* **100,** 5–28 (2004).
6. Calvo, P., Remunan-Lopez, C., Vila-Jato, J. & Alonso, M. Novel hydrophilic chitosan-polyethylene oxide nanoparticles as protein carriers. *J. Applied Polymer Science* **63,** 125–132 (1997).
7. Fan, W., Yan, W., Xu, Z. & Ni, H. Formation mechanism of monodisperse, low molecular weight chitosan nanoparticles by ionic technique. *Colloids and Surfaces B: Biointerfaces* **90,** 21–27 (2012).
8. Mertins, O. & Dimova, R. Insights on the interactions of chitosan with phospholipid vesicles. part i: Effect of polymer deprotonation. *Langmuir* **29,** 14545–14551 (2013).
9. Koukaras, E., Papadimitriou, S., Bikiaris, D. & Froudakis, G. Insight on the formation of chitosan nanoparticles through ionotropic gelation with tripolyphosphate. *Molecular Pharmaceutics* **9,** 2856–

2862 (2012).

10. Bhumkar, D. & Pokharkar, V. Studies on effect of ph on cross-linking of chitosan with sodium tripolyphosphate: A technical note. *AAPS Pharm. Sci. Tech.* **7,** E138–E143 (2006).

11. Rinaudo, M., Pavlovand, G. & Desbrieres, J. Solubilization of chitosan in strong acidic medium. *J. Polym. Analysis Characterization* **5,** 267–276 (1999).

4

CHAPTER

SYNTHESIS AND CHARACTERIZATION OF NEW THIOLATED CHITOSAN NANOPARTICLES OBTAINED BY IONIC GELATION METHOD

Reynaldo Esquivel, Josué Juárez, Mario Almada, Jaime Ibarra, and Miguel A. Valdez

Departamento de Investigación en Física, Universidad de Sonora, Rosales y Transversal, 83000 Hermosillo, SON, Mexico

ABSTRACT

We derivatized low molecular weight chitosan (LMWC) with 3-mercaptopropanoic acid (3-MPA) by a coupling reaction. The chemical modification of LMWC was characterized by Fourier transform infrared spectroscopy (FT-IR) and nuclear magnetic resonance, 1HNMR.We researched the influence of 3-MPA on the nanoparticles formation by ionic gelation method using sodium tripolyphosphate (TPP) as cross-linker reagent. In order to optimize the nanoparticles formation, we studied the effect of the pH solution and molar ratio on nanoparticles stability. Analyses of particle size, morphology, and surface charge were determined by dynamic light scattering, Atomic Force Microscopy, and zeta potential, respectively. It was found that formation of semispherical and stable nanoparticles was improved due to the chemical modification of chitosan. Optimized semispherical nanoparticles of thiolated chitosan were synthesized with the parameters (pH 4.7,molar ratios 1 : 106).

Additionally, we reported the thermodynamic profile of the nanoparticles formation determined by isothermal titration calorimetry (ITC).The aggregation process achieved to form nanoparticles of thiolated and nonmodified chitosan consisted of two stages, considering one binding site model. Gibbs free energy (ΔG) and binding constant (Ka) describe the aggregation process of thiolated chitosan/TPP, which is an initial reaction and followed by an endothermic stage.These results are promising for the possible application of these nanoparticles as nanocarriers and delivery systems.

INTRODUCTION

Nanotechnology is often used to solve problems in several areas as electronics, environment, agriculture, and biotechnology [1–4]. However, interesting properties of biopolymeric nanoparticles (BNP) have been widely explored in recent years by several research groups [5, 6]. Nowadays, BNP have been used for the development of nanocarriers and drug delivery systems by improving their biocompatibility, specificity, and charge/release mechanism. These nanoparticles have several advantages as compatibility with the biological environment, controlled release of biomolecules, and improvement in cellular penetration. It is known that surface charge of the nanostructures plays an important role in adherence and cellular uptake mechanism through electrostatics interactions with cellular membranes. For this reason, several research groups have developed versatile methodologies to elaborate nanostructured materials that have a potential application in medical areas as therapeutic agent, contrast agent, among others [7–9]. To achieve this goal, it is important to use biocompatible polymers, for example, polyethylene glycol (PEG) [10], chitosan [11], and poly (lactic-co-glycolic acid) (PLGA) [12]. Nowadays, it is a challenge to design monodisperse and stable polymeric particles useful in certain applications, including cancer therapy and gene therapy.

Chitosan is a cationic linear polysaccharide obtained by segmental alkaline deacetylation of chitin. The structure of chitosan consists of units of N-acetylglucosamine and glucosamine

[13, 14]. This polymer has been extensively studied due to the advantageous physicochemical properties such as high biocompatibility,

nontoxicity, biodegradability, good solubility in water, and facility to be used as drug delivery system [15]. On the other hand, outstanding properties allow the use of chitosan as a material for potential medical applications [11]. Thus, numerous studies have highlighted the importance of size in nanoparticles synthetized from LMW chitosan by ionic gelation method; they showed the advantageous properties of nanosized particles.

Ionic gelation is based on electrostatic interactions between negatively charged functional groups that belong to the cross-linker agent. Frequently, tripolyphosphate (TPP) is used as well as positively charged amino groups of native chitosan [16]. Fan et al. [17] studied the influence of molar ratio Ch/TPP, pH, and temperature on formation mechanism of nanoparticles of LMW chitosan. They reported NPs with hydrodynamic diameter of 138 nm and ξ-Potential value of 35mV. The physicochemical properties depend highly on reaction parameters such as concentration of crosslinker agent, pH, and ionic strength. However, they obtained microparticles at high amount of molecules of cross-linker (TPP). This problem can be overcome by structural modification of chitosan with functional groups that improve ionic interaction at the cross-linking process.

On the other hand, chemical modification of chitosan can be useful to enhance its physicochemical properties. Primary amino group (-NH2) of the chitosan molecule is relatively easy to modify due to the electron-donating effect with various functional groups such as carboxylic acids and other reactive groups [18–20]. Several chemical routes have been developed to prepare derivative chitosan by nucleophilic reaction allowing the formation of imide (aldehyde reaction) and the corresponding amide with acyl reagent as carboxylic acids. Emmanuel et al. [21] reported the influence of structuralmodification by hydrophobization on chitosan backbone.They conjugated this biopolymer with different hydrocarbon chain length (twelve carbon chains) and researched the preparation of nanoparticles by ionic gelation method.

A further modification of positively charged amine group using thiol segments is widely used in order to enhance the bulk polymer properties [22, 23]. Typical chitosan modification with thiol groups includes covalent attachment between the amine group and the mercaptocarboxylic acid by means of reaction intermediate from carbodiimide (EDAC) and Nhydroxysuccinimide (NHS) [24–27]. For instance, Ko et al.

[28] synthesized a novel thiolated chitosan by coupling reaction of 3-mercaptopropionic acid, LMWC/EDAC complex.

Thiolated chitosan has several advantageous features compared to native chitosan such as enhancement of solubility at low degree of substitution and mucoadhesive and cellular permeation properties.The chemicalmodification in certain chitosan chains allows the formation of internal and xternal bisulfide bonds.These interactions display enhancement of cohesive properties in the process of cross-linking; in this sense, the insertion of thiol groups results in improved internal interactions and high stability of microparticles [29– 31].

Zhu et al. [32] developed a new two-step method to synthesize thiolated low molecular weight chitosan with a high degree of substitution (Figure 1). They prepared spherical particles by ionotropic gelation method, using tripolyphosphate pentabasic as cross-linking reagent.DLS measurements show particle size corresponding to DH = 265.7nm and ξ- Potential 29.5mV.

Isothermal titration calorimetry (ITC) is a physical technique used to determine thermodynamic properties in biomolecular reactions. Measuring absorbed or released heat of binding process allows determination of stoichiometry reaction (N), enthalpy (ΔH), binding constant (Ka), Gibbs

free energy (ΔG), and entropy (ΔS). The thermodynamic profile of chitosan has been extensively studied in recent years; interactions between chitosan-insulin [33], chitosantrastuzumab antibody [34], and chitosan-NaCl [35] indicate important variations in the thermodynamic profile. This is attributed to the affinity of ligand-macromolecule interaction. On the other hand, few works have been produced in order to research the thermodynamics involved in NP's formation mechanism using chitosan or chitosan derivatives. In this study, we synthesized thiolated chitosan and studied the preparation of nanoparticles by ionic gelation method.We also analyzed the impact of parameters including molar ratio Ch-SH/TPP and pH of the solution in order to develop nanoparticles with potential uses in biomedical application as drug nanocarriers systems. In addition, we determined the thermodynamic profile of the process on nanoparticles formation of Ch-SHSH/TPP by isothermal titration calorimetry.

EXPERIMENTAL

Materials

Low average molecular weight chitosan (100,000 g/mol) with degree of deacetylation (75–85%) and viscosity of 20–300 cP from Sigma-Aldrich. 3-Mercaptopropionic acid (HSCH2CH2CO2H) g/mol, N-(3-dimethylaminopropyl)- N-ethylcarbodiimide hydrochloride EDAC, (C8H17N3·HCl), N-hydroxysuccinimide NHS (C4H5NO3), sodium hydroxide (NaOH), acetic acid (CH3CO2H), pentasodium tripolyphosphate TPP (Na5P3O10); all chemical reagents were purchased from Sigma-Aldrich Co., USA, and used as received. H2O was filtered with an Easy Pure/Barnstead instrument with a resistivity of 17.5Ω-cm.

Methods

Synthesis of Thiolated Chitosan

The method of synthesis was adapted according to Ko et al. [28]. It includes two stages of reaction. In the first step, 0.104mmol of EDAC, 0.208mmol of 3-mercaptopropanoic acid, and 0.104mmol of NHS were poured in 100mL three-neck flask previously charged with 2mL of dimethylformamide. The mixture was kept at room temperature and under constant stirring overnight. The first step of reaction allows formation of Oacylisourea and activation of intermediate amine-reactive sulfo-NHS ester. Finally, 20mL of chitosan solution 5%(v/w) was prepared using ultrapure water and pH was adjusted at 5.5 using NaOH 1M. We added intermediate solution dropwise by syringe. The final mixture was stirred for 12 hours at room temperature. To eliminate the unbound 3-MPPAand isolate the thiolated chitosan, the raw mixture was dialyzed (Cellu/Sep membrane dialysis MWCO = 5000) using 0.01M HCl and deionized water. The product was precipitated with 50mL of ethanol and the polymer was separated by centrifugation (5,000 rpm) (3 times Allegra X-22 centrifuge, F1010 rotor). Derivative chitosan was lyophilized and then characterized by spectroscopic techniques to ensure pure chitosan.

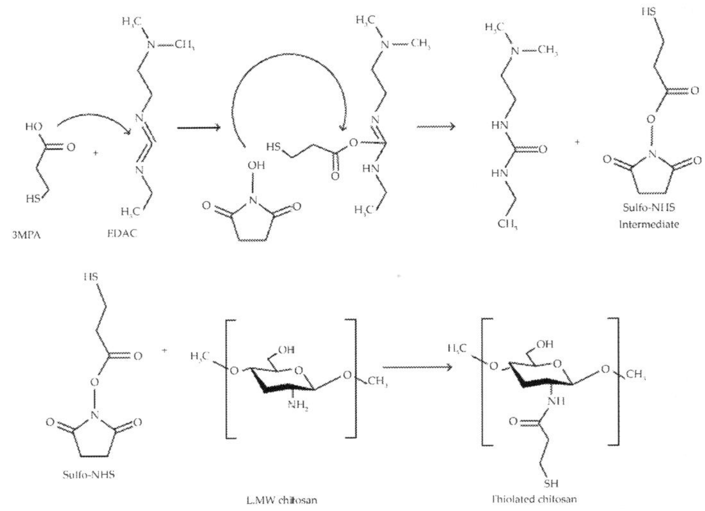

Figure 1: Reaction mechanism of thiolated chitosan.

Preparation of Thiolated Chitosan Nanoparticles

Chitosan thiolated nanoparticles were synthesized by adaptation of Zhu et al. method [32], based on the ionic gelation by cross-linking of TPP and free chains of chitosan. Synthesis of thiolated chitosan nanoparticles was as follows: thiolated chitosan (0.5mg/mL) was dissolved in aqueous diluted solution of acetic acid (2%w/w) and stirred overnight at room temperature. The pH of the solution was around 3.9; this value was adjusted to 4.7 by addition of minimal amounts of NaOH 2Msolution. After that, aqueous solution of tripolyphosphate was prepared at concentration of 0.5mg/mL using ultrapure water, and then 3.3mL of this solution was stored and cooled in a fridge at 0–2∘C for 4 hours. The solution of thiolated chitosan was placed in preheated water bath at 60∘C for 10 minutes under constant stirring. Then this solution was transferred to an ice bath and, immediately, TPP solution was added to the mixture and stirred for ten minutes. The ice bath was removed and the obtained opalescent suspension was further stirred for fifteen minutes at room temperature. The synthesized nanoparticles were centrifuged at 15000 rpm by 30 minutes and resuspended twice in ultrapure water for physicochemical characterizations. We used the same procedure with different molar ratio Ch-SH/TPP (1 : 68 to1 : 109).

Characterization

Fourier Transform Infrared Spectroscopy (FT-IR).

Fourier transform infrared spectra were collected from spectrometer Spectrum Two (Perkin-Elmer, USA) equipped with an accessory of a single diamond Attenuated Total Reflectance (ATR) with range 8300–350 cm−1 and interferometer with variable speed of 0.1, 0.2, 0.5, 1, 2, and 4 cm/s. IR Beam diameter is from 2 to 11 mm. The IR spectra were collected sixteen times (spectral resolution 4 cm−1) and analyzed using Spectrum software. Chitosan was analyzed before and after derivatization with thiol groups. Spectra were recorded and analyzed for signal assignation.

Nuclear Magnetic Resonance (1HNMR)

High resolution 1HNMR spectra were recorded from Bruker (Avance, 400MHz) ultra-shielded standard-bore magnet equipped with a 5mmBBFO probe and a Bruker Sample Case autosampler with 24 positions. Native and thiolated chitosan were analyzed in order to confirm structural modification in pure chitosan. All the samples were prepared dissolving 8mg of the polymer with 2% (v/v) of DCl/D2Omixture. The solution was heated at 50°C to ensure complete solubility. Results were analyzed in ACD/labs software for spectra calibration and peak assignment.

Size Measurements

Dynamic light scattering (DLS) measurements were performed in Zetasizer Nano *Z* equipment (Malvern US), using digital autocorrelator 25 ns to 8000 s, max 4000 channels with measurement range of 0.3 nm–10.0 microns, equipped with temperature control set to 0°C to 90°C ± 0.1°C, with He-Ne laser 4mW 633nm and purge system using dry air. The method used is Noninvasive Back-Scatter (NIBS) technology. The relationship between the size of a particle and its speed due to Brownian motions is defined in the Stokes-Einstein equation:

$$D = \frac{K_B T}{6\pi\eta R}, \qquad (1)$$

where D is diffusion constant, KB is Boltzmann's constant, T is the absolute temperature, η is dynamic viscosity, and R is hydrodynamic radius. The analysis was carried out in triplicate in diluted suspensions of nanoparticles.

Zeta Potential Determination

Determination of zeta potential was carried out in Zetasizer Nano Z with light source He-Ne laser 633 nm, max 4mW with measurement accuracy 0.12 μm·cm/V·s for aqueous systems using NIST SRM1980 standard reference material. The Zetasizer Nano series calculates the zeta potential by determining the electrophoretic mobility and then applying the Henry equation:

$$U_E = \frac{2\varepsilon Z f\,(ka)}{3\eta}, \qquad (2)$$

where UE is the electrophoretic mobility, ε is the dielectric constant, Z is the zeta potential, η is the absolute zero-shear viscosity of the medium, $f(ka)$ is the Henry function, and ka is a measure of the ratio of the particle radius to the Debye length.The velocity of a particle in an electric field is commonly referred to as its electrophoretic mobility. The samplewas analyzed in quartz cuvettes using a small amount of colloidal suspension. Zeta potential and standard deviation were reported.

Atomic Force Microscopy (AFM)

AFM imaging was performed by an AFM instrument (JSPM-4210, JEOL) in no contact mode. A cantilever (OMCL-AC160TS, Olympus) with resonant frequency of 300 kHz and spring constant of 42 N/mwas used. Resuspended thiolated and native nanoparticles were diluted in ultrapure water 1 : 100 and poured in a new cleaved mica surface. Images were treated and analyzed in order to elucidate morphology and size of particles.

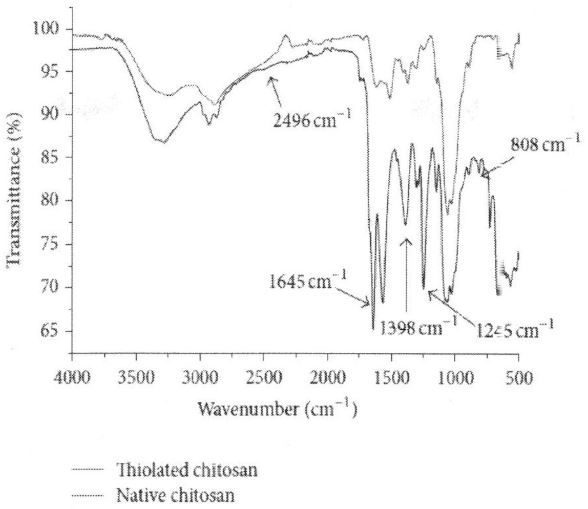

Figure 2: FT-IR spectra of thiolated chitosan and native chitosan.

Isothermal Titration Calorimetry (ITC)

Isothermal titration was developed by MicroCal VP-ITC with automatic injection syringe system (300 μL max). Isothermal titration calorimetry (ITC) measures heat change in an ITC experiment and enables accurate determination of binding constants (KD), interaction stoichiometry (N), enthalpy (ΔH), and entropy (ΔS), providing a complete thermodynamic profile of the molecular interactions in a single experiment. Experiments of titration for the nanoparticles synthesis from native and thiolated chitosans were performed as follows: sample cell was filled with 1.4584mL of solution 1.6 . 10−3 mM of low molecular weight chitosan at pH 4.7; acetic acid 30mM solution at the same pH was charged in a reference cell. All solutions were previously degasified under vacuum. Sampler micropipette was loaded with TPP 0.98mMsolution, using ultra-pure water (18.3 μ-Ω-cm).The experiment was performed with 26 injections of 10 μL of TPP solution every 300 seconds and an initial injection of 2 μL. Raw data were treated with MicroCal ORIGIN version 7.0 software for background subtraction and mathematical modeling.

RESULTS AND DISCUSSION

Spectroscopy

Fourier Transform Infrared (FT-IR)

Comparative FTIR spectra of native LMWC and thiolated chitosan Ch-SH are shown in Figure 2. Results of the analysis indicate the successful modification of native chitosan selectively on the amine groups, by the reaction between the primary amine groups of chitosan and the carboxylic groups of themercaptopropionic acid resulting in amide group. The presence of the peak at 1645 cm−1 (C=O stretching amides) and deformation of the signal at 3225 cm−1 (-NH stretching amide) proves the formation of (C-NH) amide bond.Thepeak at 2496 cm−1 due to (-SH) stretching confirms the presence of thiols as terminal group linked to the chitosan molecule. The presence of the peaks at 808 cm−1 (S-S bisulfide bond) and 1245 cm−1 (CSH stretching) inmodified chitosan confirms the existence of mercaptans since such peaks are not seen in native chitosan spectra.

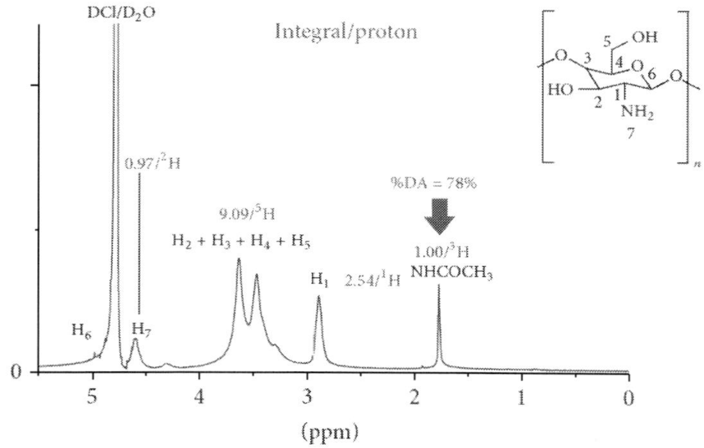

(a)

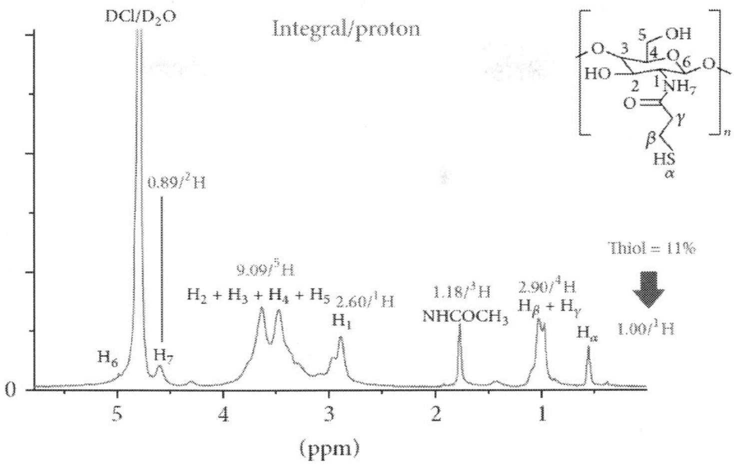

(b)

Figure 3: 1HNMR spectra of native (a) and thiolated (b) chitosan in DCl/D2O.

Compared to the results already published by Zhu et al. [32], they reported modification of native chitosan using thioglycolic acid; structural characterization by FTIR spectra shows typical peaks of mercaptans at 2680 cm−1 (SH stretching), 1250 cm−1 corresponding to the vibration of C-SH bond, and small peak at 2923 cm−1 associated with stretching of C-H groups. FT-IR analysis confirms the success of modification of native chitosan by addition reaction and a complementary analysis of derivative compound was carried out by 1H protonmagnetic resonance (1HNMR).

1H Proton Magnetic Resonance (1HNMR)

Functionalized and native LMWC were characterized by 1HNMR spectroscopy under the same conditions; corresponding 1HNMR spectra in DCl 10mM of thiolated and native LMWC are shown in Figure 3. In the NMR spectrum of the thiolated chitosan, there still appeared some signals similar to those native chitosan.The appearance of signals at 1.92ppm (NHCOCH3), 2.98 ppm (H1) due to the protons -CH-, and 3.6 to 3.8 ppm (H2, H3, H4, and H5) due to CH protons confirms the presence of the glucosamide moiety. Decrease of signal at 4.6 ppm (H7, NH2) and new peak at 0.52 ppm (Hα) (-CH2-SH) are in accordance with the formation of thiolated chitosan.The signal at 0.92 (Hγ) to 1.10ppm (Hβ) corresponds to protons of the new alkyl chain (-CH2). Han et al. [36] synthesized thiolated derivative chitosan where structural characterization by 1HNMR shows a peak at 2.1 ppm that corresponds to the protons of -CH2- of the new side chain; a small signal was observed at 4.6 ppm due to the remaining protons of amine group (-NH2-) after the derivatization of native chitosan. Intense peak at 4.8 ppm was related to the proton H1 of hydroxyl group.We consider that a high degree of derivatization could lead to insoluble systems due to the hydrocarbon chains. Thus, the 1HNMR spectra proved the successful attachment of mercaptopropionic acid to native LMW chitosan and also confirmed the presence of free thiol groups.

Quantification of Chitosan Transformation

Consider the following equation:

$$\%\text{Yield} = \left(\frac{A_{\alpha}}{A_{\text{H}_2-\text{H}_5}} \right) \times 100, \qquad (3)$$

where $A\alpha$ is area of signal SH and AH2–H5 is area of signal H2–H5.

Nuclear magnetic resonance of proton (1HNMR) was employed in quantitative analysis and structural determination of modification of native chitosan. The degree of modification of native chitosan with thiol groups (SH) was calculated from the area of the signal Hα (ASH) and the sum of the areas from signals H2 to H5 (AH2–H5) at the 1HNMR spectra (1) [37]. The amount of free thiol groups was determined by 1HNMR, and this value reached 11% of substitution on amine groups.The lowdegree ofmodification of the native chitosan avoids the insolubility problems of modification above 40%.

Study of Ch-SH/TPPMolar Ratio on NP's Formation

Effect on the Size of Nanoparticles.

We researched the influence of TPP at different molar ratio with respect to thiolated Ch-SH (1 : 71 to 1 : 114 Ch-SH/TPP 0.5mg/mL) on the particle size.The obtained results are presented in Table 1. From DLS data (Figure 4), we noticed an evident decrease of hydrodynamic diameter of nanoparticles by increasing the amount of TPP. This fact demonstrates the influence of the amount of TPP units on the NP's size. The first decrement of particle size at molar ratio 1 : 71 to 1 : 86 (Table 1) was from 195 nm to 168 nm; consecutively the size of nanoparticles decreased to 142nm. When the molar ratio of Ch-SH/TPP was 1:71, the result was a slight opalescent suspension, indicating that the density of TPP is barely enough for crosslinking.

Table 1: ξ-Potential and size, effect of Ch-SH/TPP molar ratio concentration 0.5 mg/mL.

Sample	Molar ratio	pH	ξ-Potential (mV)	Size (nm)	PDI
NPT_1	1:114	4.7	30	142	0.16
NPT_2	1:106	4.7	34	137	0.16
NPT_3	1:100	4.7	31	162	0.14
NPT_4	1:94	4.7	39	160	0.18
NPT_5	1:86	4.7	40	168	0.18
NPT_6	1:77	4.7	39	200	0.18
NPT_7	1:71	4.7	40	195	0.20

This behavior was confirmed by AFM. Chitosan chains have high quantity of reactive amine groups, which provides potential cross-linking sites. Thereby, the molar ratio of Ch-SH/TPP is related with the crosslinking degree, but it is limited by the aggregation phenomena at high concentration of the polyanion. In this sense, when the amount of TPP molecules of Ch-SH/TPP reached 1 : 114, a decrease in the particle size was observed due to a higher amount of TPP molecules, which promote the nanoparticle formation and provide structural stability of chitosan nanoparticles by interaction between the TPP molecules and positively charged free chitosan chains, shrinking chitosan NPs. Fan et al. [17] prepared nanoparticles from native chitosan for a specific range from 1 : 86 to 1 : 90 of molar ratio of TPP/Ch; their results showed that above molar ratio of 1 : 96, large microparticles are produced, thus limiting the repulsion forces between the particles leading to the flocculation and subsequent precipitation. Modulation of particle size prepared frommodified chitosan is an important parameter for cellular internalization. Cells can internalize particles under 1 μm in diameter by different pathways. For example, particles as large as 500nm could be internalized via an energetic dependant mechanism. In contrast, particles under 200 nm in size can lead to a nonenergetic endocytosis pathway. The modulation of nanoparticles of thiolated chitosan (142nm to 195nm) is an advantageous factor in order to design nanostructures for specific application. In this work, we elaborated nanoparticles using thiolated chitosan even at high concentration of TPP molecules.

Thereby, at molar ratio 1 : 94, unimodal nanoparticles of thiolated chitosanwere formed.The size of these particles was modulated from 142nm at molar ratio 1 : 114 (Ch-SH/TPP) to 195 nm prepared with molar ratio 1 : 71 (Ch-SH/TPP). We associate the stability of thiolated chitosan NPS at highest amount of TPP with the formation of external and internal disulfide bonds; in this manner formation on nanoparticles is stabilized by stronger interactions from disulfide bonds [38, 39].

Effect on ξ-Potential

In this section, we analyzed the effect of different molar ratio Ch-SH/TPP (1 : 71 to 1 : 114 Ch-SH/TPP 0.5mg/mL) on the zeta potential of nanoparticles. According to the results shown in Table 1, we observed an important decrease of zetapotentialby increasing the amount of TPP molecules. Zeta potential decreased proportionally to the increase of TPP molecules from 40mV to 30mV; this is due to the charge neutralization reaction between amine groups of chitosan and free chains and negative charges of TPP. For all cases, zeta potential values were positive and relatively high (above 30mV). Particles showed good stability at room temperature.

Zeta potential is related to the repulsive or attractive interactions between adjacent particles. For biomedical application, the modulation of charge is important in order to design nanodevice able to interact with positive or negative barriers or membranes.

Effect on Morphology

An analysis of morphology was carried out by Atomic Force Microscopy (AFM). Figure 5 shows AFM micrographs for NPs with different molar ratio. In the set of micrographs, it is possible to observe the morphological evolution of thiolated chitosan nanoparticles as function of TPP concentration. Figure 5(a) (Ch-SH/TPP 1 : 114) shows semispherical nanoparticles with an average size of 80 nm. In this sense, it is known that intra- and intermolecular interaction between TPP and chitosan are responsible for the ionic gelation even at a low concentration of TPP. Figure 5(b) presents monodisperse particles with sizes between 140nm and 200nm (Figure 6) corresponding to the DLS measurements. In Figure 5(c), it can be seen that at this concentration of anionic cross-linker, morphology was semispherical, maybe due to the decrease of cross-linking degree of the chitosan-TPP reaction. It is important that when the molar ratio of Ch-SH/TPP was 1 : 94, Figure 5(d), the population of nanoparticles were again polydisperse. We observed that a high quantity of molecules of TPP could change the structural conformation of chitosan chains and significantly reduce the solubility of the chitosan [40, 41], thus favoring aggregation. We reduced the amount of TPP molecules (molar ratio 1 : 77 andmolar ratio 1 : 71); under this condition, we obtained nanoparticles with size between 120 to 150 nm.The decrease of particle size is compared with the results obtained by Weng F. et al. (2012). They determined the best condition of molar ratio of Ch/TPP (1 :90) to obtain chitosan nanoparticles. At molar relation Ch/TPP 1 : 96, the size increased to 240 nm; probably the excessive concentration of TPP limited the intramolecular cross-linking. An overall conclusion of our results is that it is possible to modulate size and charge of nanoparticles between 100 nm and 200 nm, and between 30mV and 40mV, respectively. A greater dependency of the formulation parameters was observed in themorphology and population of nanoparticles.

Effect of pH ofThiolated Chitosan Solution on NP's Formation

Effect on Particle Size

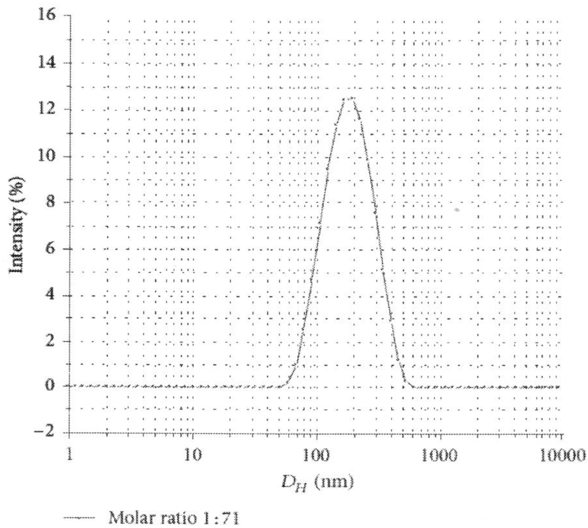

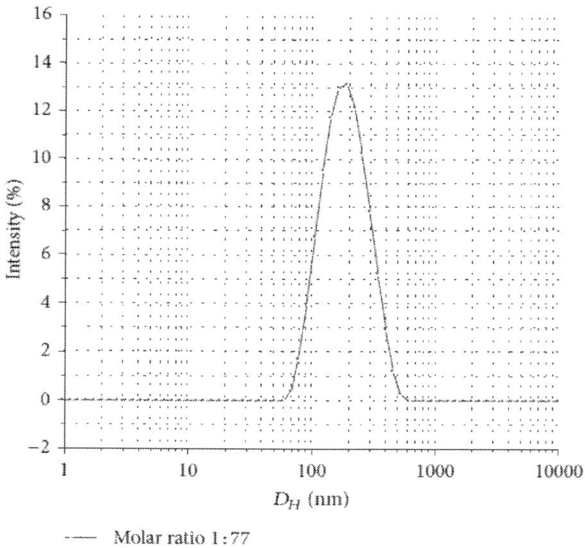

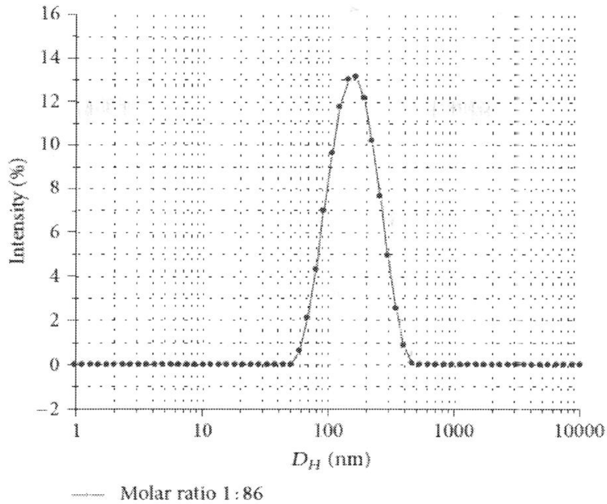
Intensity (%)
D_H (nm)
Molar ratio 1:86

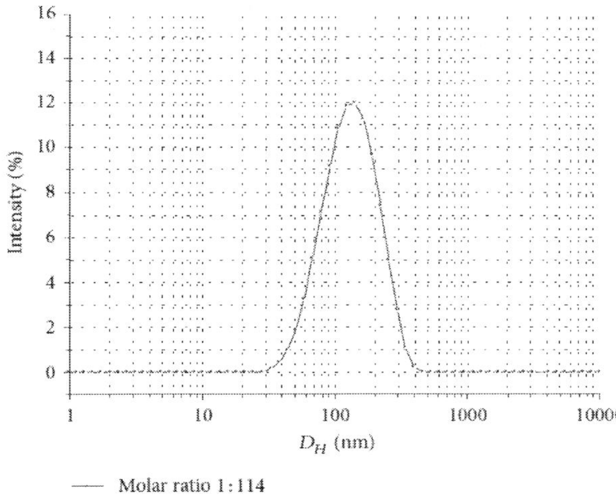
Intensity (%)
D_H (nm)
Molar ratio 1:114

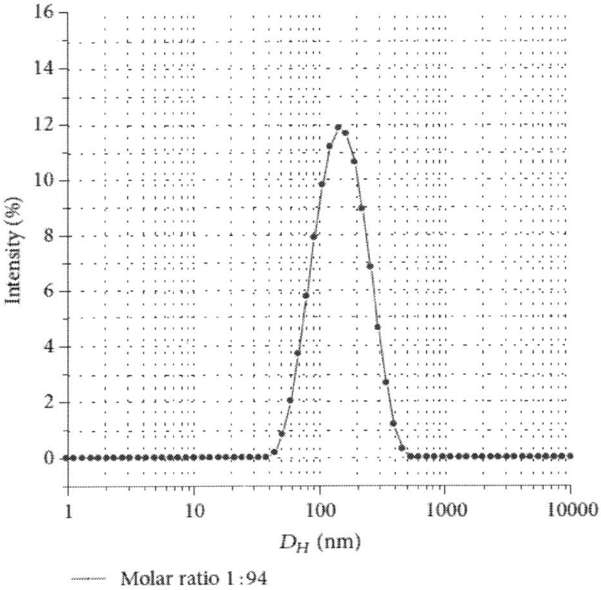

Intensity (%)
16
14
12
10
8
6
4
2
0
1
10
100
1000
10000
D_H (nm)
Molar ratio 1 : 94

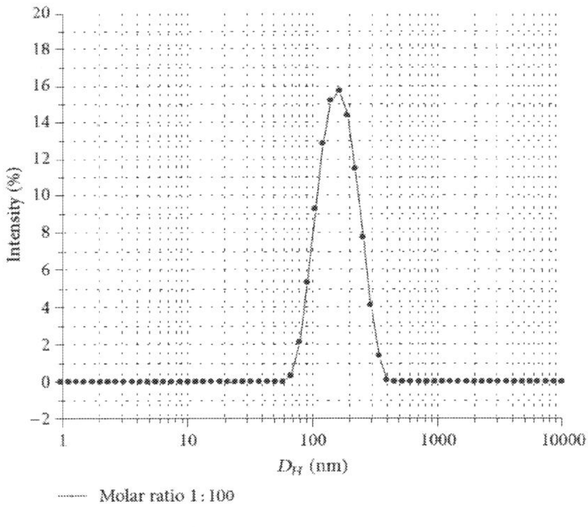

Intensity (%)
20
18
16
14
12
10
8
6
4
2
0
−2
1
10
100
1000
10000
D_H (nm)
Molar ratio 1 : 100

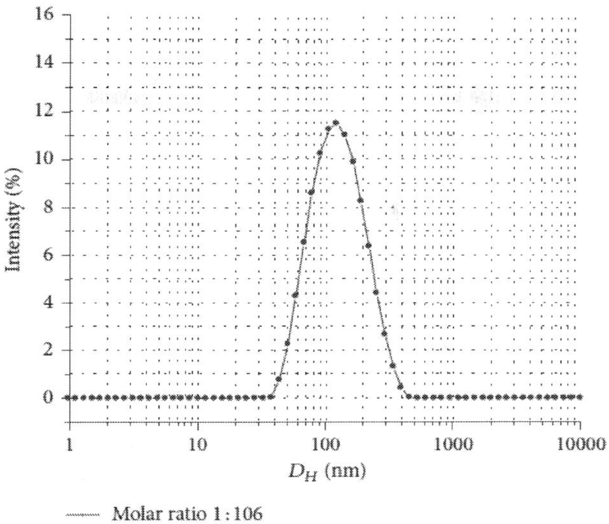

Figure 4: Particle size distribution of chitosan SH/TPP nanoparticles prepared with different molar ratio.

TABLE 2: ξ-Potential and size, effect of pH.

Sample	Molar ratio	pH	ξ-Potential mV	Size nm	PDI
$NPpH_1$	1:94	4	38	153	0.11
$NPpH_2$	1:94	4.3	38	184	0.11
$NPpH_3$	1:94	4.5	42	189	0.07
$NPpH_4$	1:94	4.7	39	162	0.18
$NPpH_5$	1:94	5	40	215	0.21
$NPpH_6$	1:94	5.3	33	220	0.15

In order to determine the feasibility to prepare thiolated chitosan nanoparticles by ionic gelation, the pH of chitosan solution was varied from 4 to 5.3. Chitosan is a weak polyelectrolyte with pKa near to (6.5) due to the amine groups and insoluble at alkaline and neutral pH. At acid medium pH, the primary amine can be protonated by hydrogen ions and thus acquires positive charge density. Our study was based on controlling the acid conditions of the solution; in this way, the dependence of the degree of protonation of amine groups of chitosan on the conformational arrangements of free chains is analyzed. To assess the effect of pH on NP's formation, we selected six different pH values (4, 4.3, 4.5, 4.7, 5.0, and

5.3), and measured zeta potential, hydrodynamic diameter and observed morphology. Figure 7 shows the DLS profile for each experiment, we observed monodisperse curve, and PDI values below 25%. Table 2 shows that hydrodynamic diameter increases significantly as pH takes a value of 5.3.At the highest pH measured (5.3), microaggregation phenomenon occurred due to nearness pKa of the chitosan (pKa = 6.3) [42]. However, at low pHconditions (4 to 4.5) the amine groups are protonated (NH3+) and favor the extension of chitosan free chains. On the other hand, if pH increases near to isoelectric point, a deprotonation process could occur and lead to microaggregation by decreasing repulsion forces between particles. In the case of pH (4.7), amino groups of chitosan chains were further positively charged and increased electrostatic repulsion between chitosan free chains. Therefore, higher energy is needed to overcome the repulsion forces leading to formation of smaller diameter NPs.

Effect on ξ-Potential

Table 2 summarizes the influence of pH on physicochemical properties of thiolated chitosan nanoparticles. We observe zeta potential of samples at lowpH from 4 to 5 which were almost constant with slight variations, except for sample at pH 5.3 (30mV) being that, at this pH, a higher degree of deprotonation of (NH3+) could happen and drop down positive charges by neutralization process resulting in unstable particles and aggregation phenomenon.

Effect on Morphology

Morphology of thiolated chitosan NPS at different conditions of pH (4, 4.3, 4.5, 4.7, 5, and 5.3) was determined by Atomic Force Microscopy. In Figure 8, we observe the morphological evolution of nanoparticles prepared from thiolated chitosan. Figure 8(a) shows particles at pH 4; the image exhibits small and monodisperse particles with size from 150nm to 160 nm (Figure 9). The shape consists of semispherical arrangements caused by well dilution of chitosan chains at higher hydrogen ions concentration, favoring prearrangement of protonated chitosan free chains, leading to formation of semispherical shapes and homogeneous population of NPs. On the contrary, when pH was increased to 4.3 and 4.5 (Figures 8(b) and 8(c)) we observed partial microaggregates which consist of smaller particles. The size of internal particles was around

140 nm; morphology was irregular. Decrease in particle homogeneity suggests the influence of concentration of hydrogen ions on protonation degree of chitosan amine. In addition, it is interesting that the evolution of microaggregates at pH 4.3 more packed structure

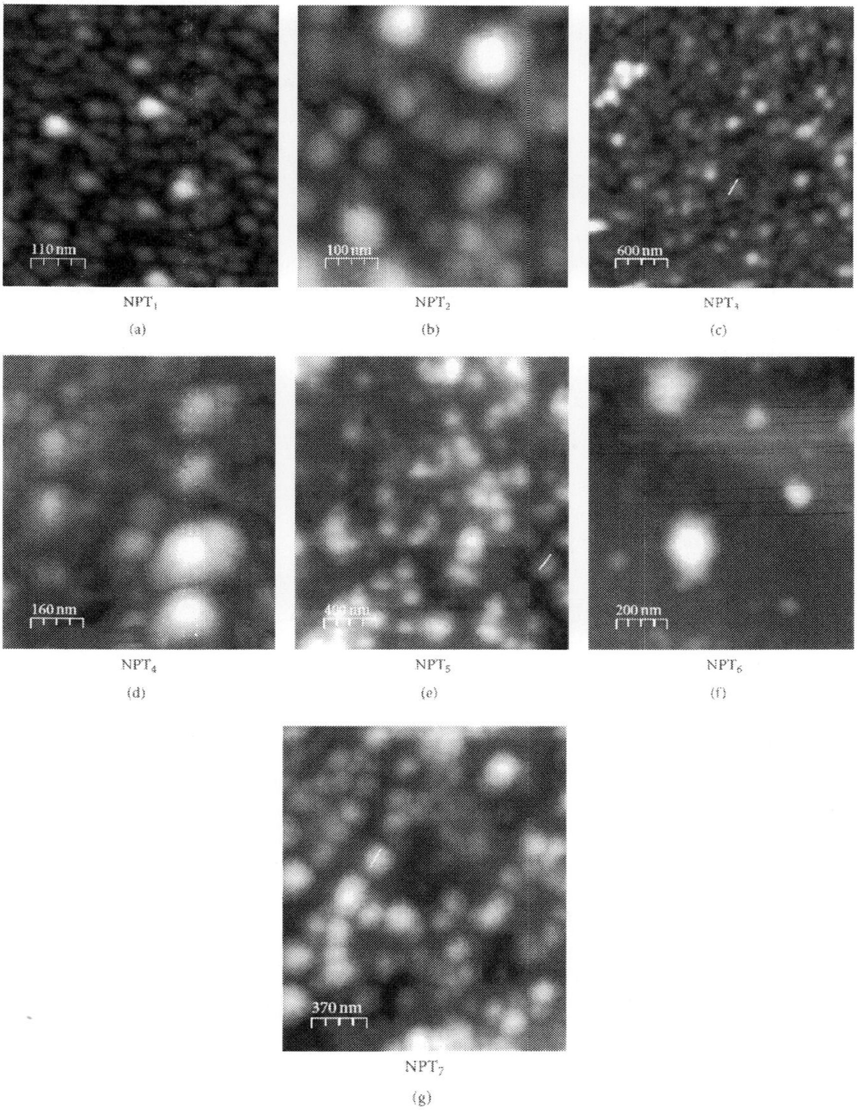

Figure 5: AFM of shape evolution of thiolated chitosan nanoparticles NPT1 (1 : 114), NPT2 (1 : 106), NPT3 (1 : 100), NPT4 (1 : 94), NPT5 (1 : 86), NPT6 (1 : 77), and NPT7 (1 : 71).

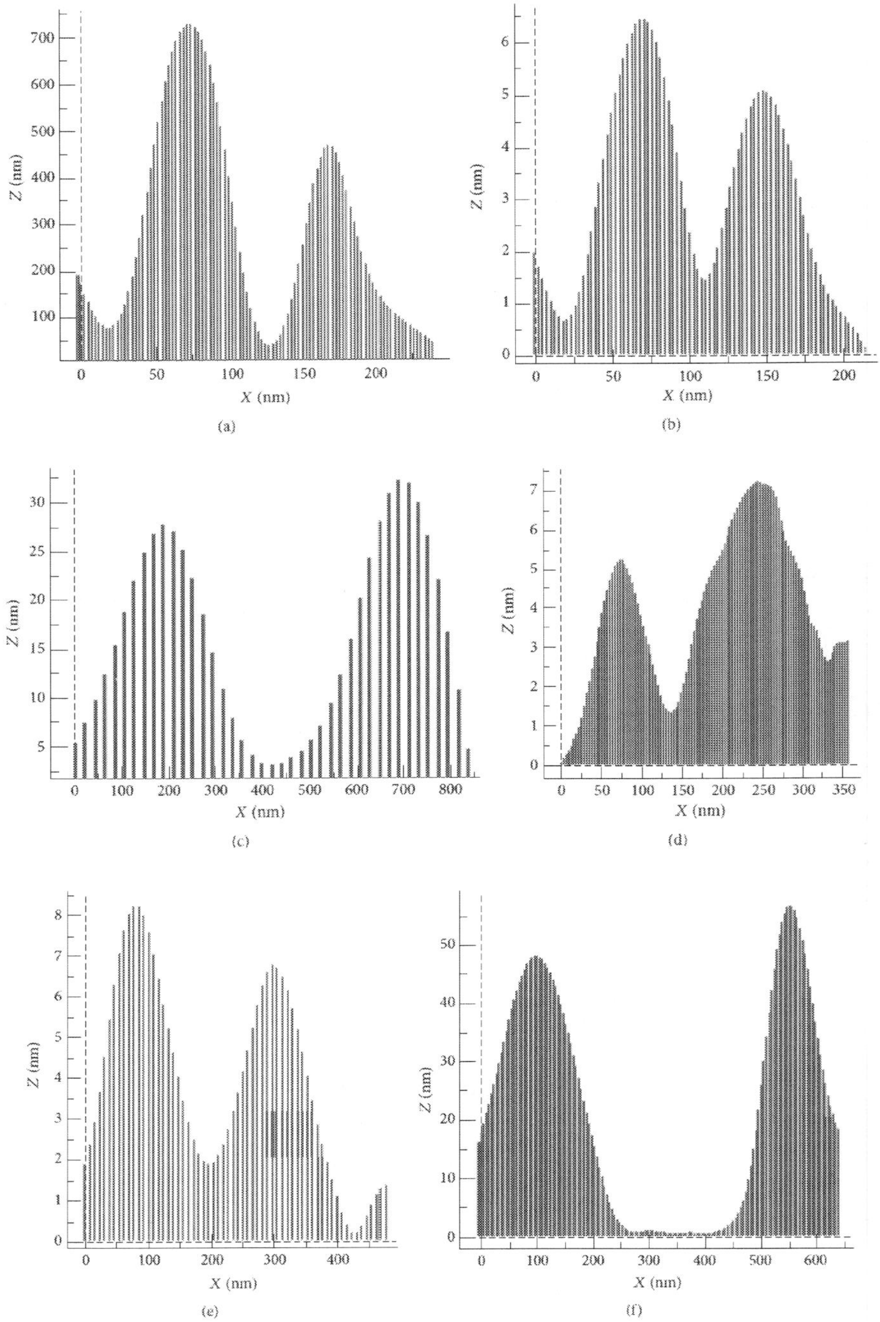
Z (nm)
X (nm)
(a)
(b)
(c)
(d)
(e)
(f)

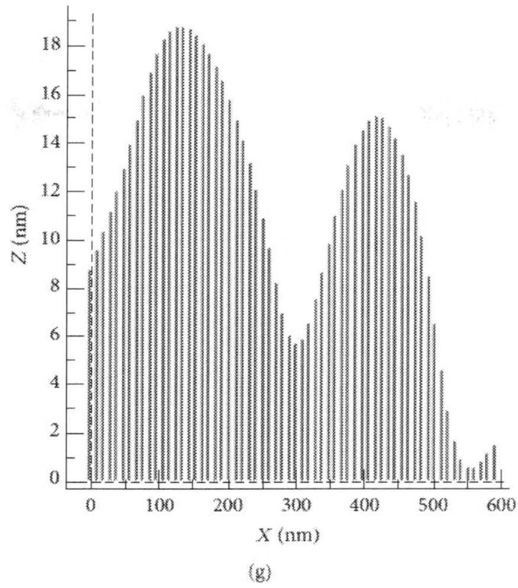

Figure 6: Profile curve of thiolated chitosan nanoparticles NPT1 (1 : 114), NPT2 (1 : 106), NPT3 (1 : 100), NPT4 (1 :94), NPT5 (1 : 86), NPT6 (1 : 77), and NPT7 (1 : 71).

Table 3:Thermodynamic profile of thiolated and native chitosan.

Sample	Gibbs energy (ΔG) cal∗mol	Enthalpy (ΔH) cal∗mol	Binding (K_a) M^{-1}	Entropy (ΔS) cal∗K^{-1}∗mol^{-1}	N (active sites) Sites
Native Ch (Phase 1)	−8621.78	−7843	2.11×10^6	2.64	20
Native Ch (Phase 2)	−8527	1.70×10^4	1.8×10^6	88.7	35
Thiolated Ch (Phase 1)	−8938	−6750	3.6×10^6	7.41	13
Thiolated Ch (Phase 2)	−8852	1.79×10^4	3.11×10^6	86	23

ΔG Gibbs free energy, ΔH enthalpy change, and ΔS entropy.

and pH 4.5 particles are further apart; this is attributed to the decrease of ionic strength limiting size of micro-aggregates at highest hydrogen ion concentration. AFM images of NPs at pH 4.7 and 5 are illustrated in Figures 8(d) and 8(e); increasing further pH condition leads to both formation of smaller particles supported by higher homogeneity and more compacted structure. At this stage, stability of the particles depends on the repulsion forces, each particle which is related to ionic strength of the environment. In these cases, an important amount of positive charges of chitosan still present in the beginning stage of aggregation will favor formation of semispherical arrangement with smaller size (160nm to 180 nm).

AFM suggest that nanoparticles are formed at pH over 4.7 to 5 favoring the electrostatic and conformational process. Physicochemical properties confirm appropriate particle size, high stability, and appropriate morphology for potentially application as for drug delivery system for target therapy.

Calorimetric Analysis of Ch-SH/TPP Binding

Nanoparticles formation of thiolated chitosan promoted by the crosslinking of TPP was researched using isothermal titration calorimetry (ITC). This technique allows determination of the mechanism of the interaction in biomolecular process. Aiming to understand the interactions Ch-SH-TPP, we analyzed data by fitting in a mathematical model of the optimized conditions, Ch-SH/TPP 1 : 106 and pH = 4.7 (NPT2). According to data analyzed of Ch-SH-TPP and Ch-

TPP, the process of nanoparticle formation by TPP crosslinking can be adjusted to an aggregation process developed in two stages, assuming that chitosan macromolecules show single binding site. For noncooperative model with single binding site, enthalpy usually keeps constant below saturation point [35, 41]. Figure 10 describes the evolution of Ch-TPP and Ch-SH-TPP nanoparticle formation and Table 3 shows ΔH, ΔS, and ΔG corresponding values. The first stage is characterized by an exothermic process, while the second stage is entropic process; however, both processes were spontaneous. The exothermic process is developed by electrostatic interactions between TPPmolecules with amine groups of Ch-TPP and Ch-SH-TPP chains. This electrostatic association allows expanded arrangements of macromolecule chains as the TPP molecules are attached to Ch-TPP or Ch-SH-TPP by attractive effect.

At this first stage, once amine groups in the Ch-TPP and Ch-SH-TPP are saturated by the anionic linker, chitosan molecules adopt a structural conformation that allow the formation of nanoparticles. This phenomenon occurs at the end of the first stage; this stage is favored by enthalpy. According to the mechanism involved in the evolution of cross-linking process to native chitosan-TPP (Figure 11) and thiolated chitosan-TPP, we observed exothermic increment

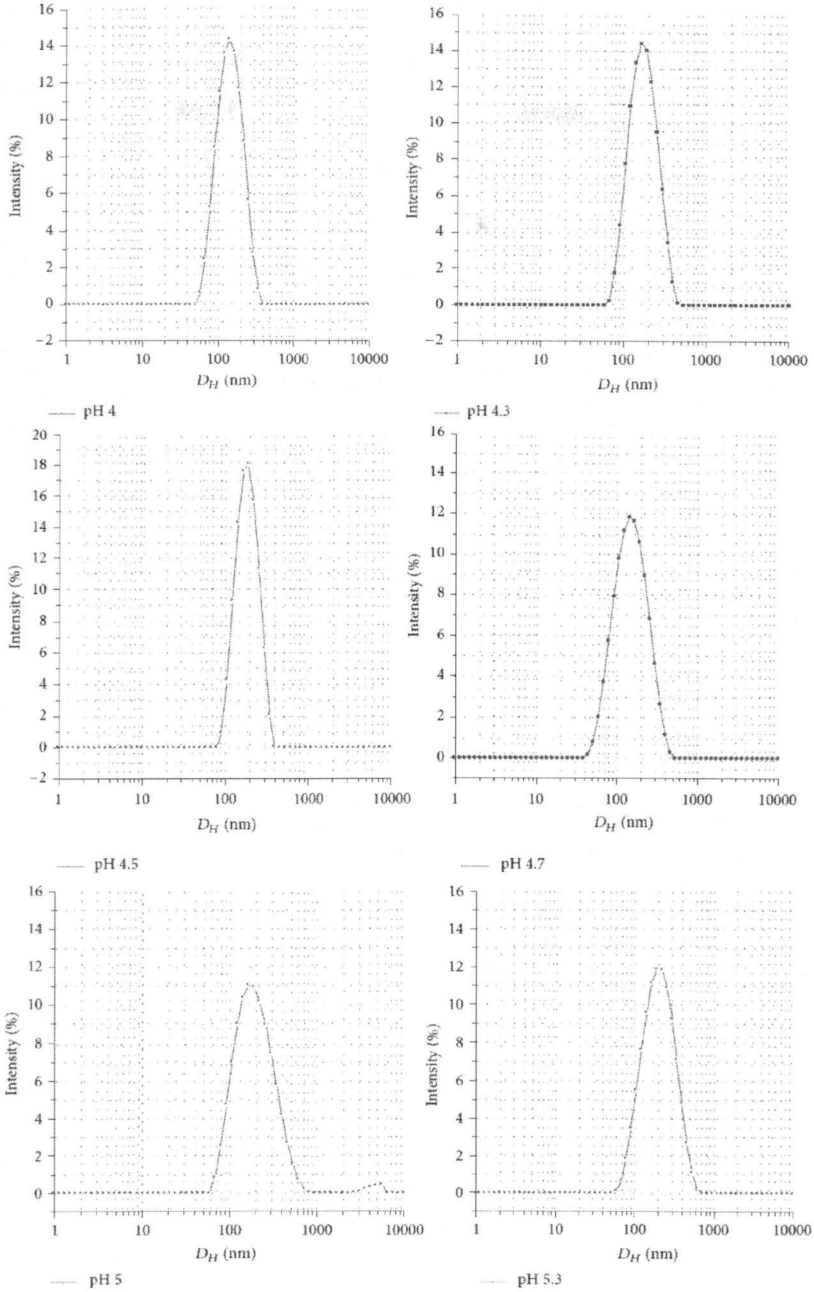

Figure 7: Particle size distribution of chitosan SH/TPP nanoparticles prepared at pH 4, 4.3, 4.5, 4.7, 5, and 5.3.

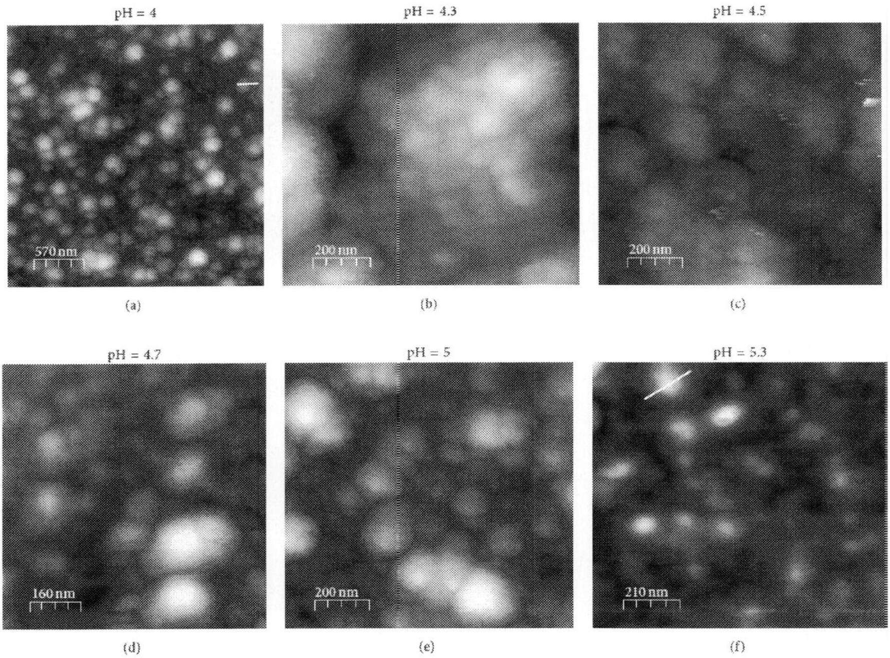

Figure 8: AFM of morphology evolution of thiolated chitosan nanoparticles

below the initial amounts of TPP; these changes can be interpreted as the first process binding. In the case of thiolated chitosan/TPP interaction, Figure 10 highlights exothermic changes at low concentration of TPP, which can be associated with initial conformational arrangements of chitosan free chains in solution from semicoil state to extended conformation.

This mechanism allows the later attachment of TPP molecules by ionic interaction between positive charges of chitosan and negative from TPP. According to thermodynamics of the first phase, the structural arrangement of thiolated chitosan chains occurs as a spontaneous process ($\Delta G = -8938$ cal/mol); the equilibrium favors the expanded conformation of chitosan chains, probably favored by the concentration of ligand and the temperature of the initial solution.

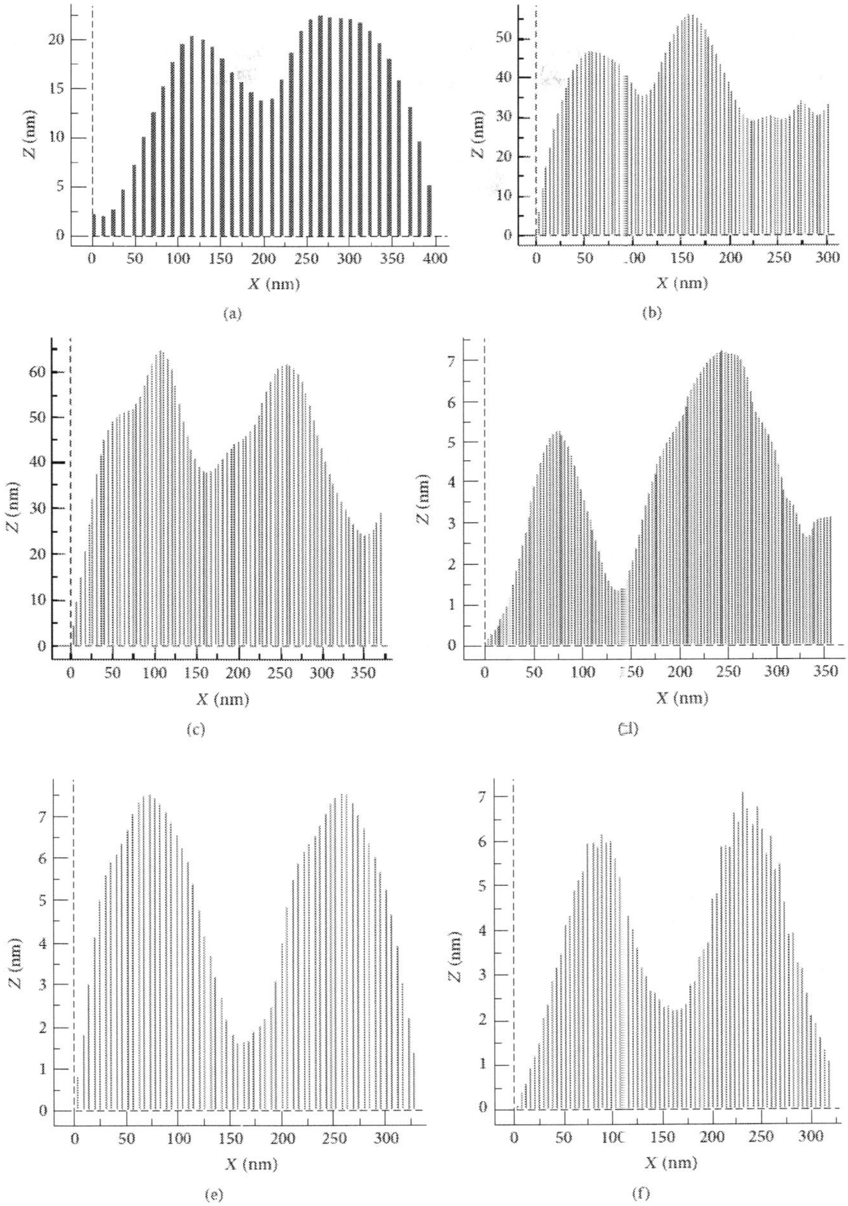

Figure 9: Profile curve of thiolated chitosan nanoparticles of Ch-SH/TPP nanoparticles prepared at pH 4, 4.3, 4.5, 4.7, 5, and 5.3.

In a direct comparison between thiolated and unmodified chitosan, Gibbs free energy reflects a higher stability in the case of thiolated chitosan by a quantitative difference of 317 cal∗mol. This change indicates a highest stability caused by interaction of disulfide bonds. On the other hand, ITC results indicate an increment of entropy change (ΔS), (7.41 cal∗K−1∗mol−1 Ch-SH-TPP), and (2.64 cal∗K−1∗mol−1 Ch-TPP); this means that modified chitosan is entropic driven by a molecular prearrangement of chitosan and TPP molecules. The second phase involves an endothermic mechanism.

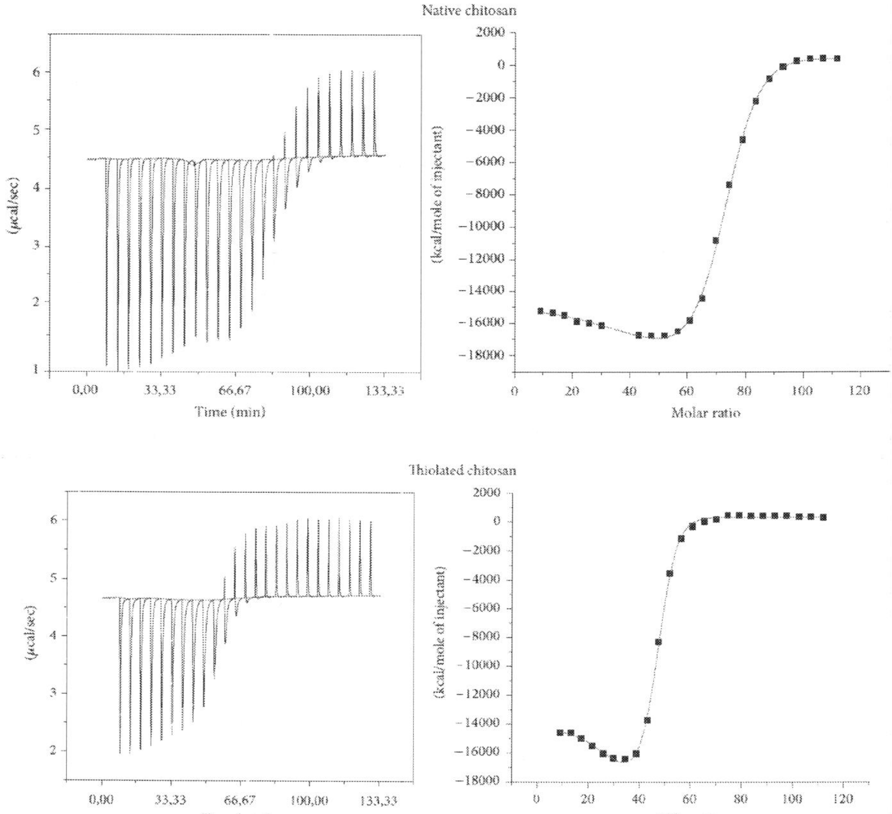

Figure 10: ITC and raw data of thiolated chitosan.

Equilibrium is favored by the linkage formation of extended chitosan occupied by TPP molecules. Additionally, the graph shows an increment of endothermic signal at the second phase, possibly due to an entropic phenomenon. This phenomenon at the second stage was initiated when chitosan active sites were occupied by TPP molecules; after that, ionic forces led to intermolecular interaction between chitosan chains. Before nanoparticles formation, water molecules are linked to backbone of chitosan chains by hydrogen bonds, the rupture of these hydrogen bonds causes energy absorption and starting the endothermic route. Entropy increment (ΔS) (88.7 cal∗K−1∗mol−1 Ch-TPP) and (ΔS) (86cal∗K−1∗mol−1 Ch-SH-TPP) involves a spontaneous structural conformation of chitosan chain. Gibbs free energy is related with the endothermic signal at higher amount of TPP, we describe this second step as a spontaneous process (Gibbs free energy −8852 cal∗mol), and this value suggests that the reaction favors Ch-SH-TPP cross-linking process.

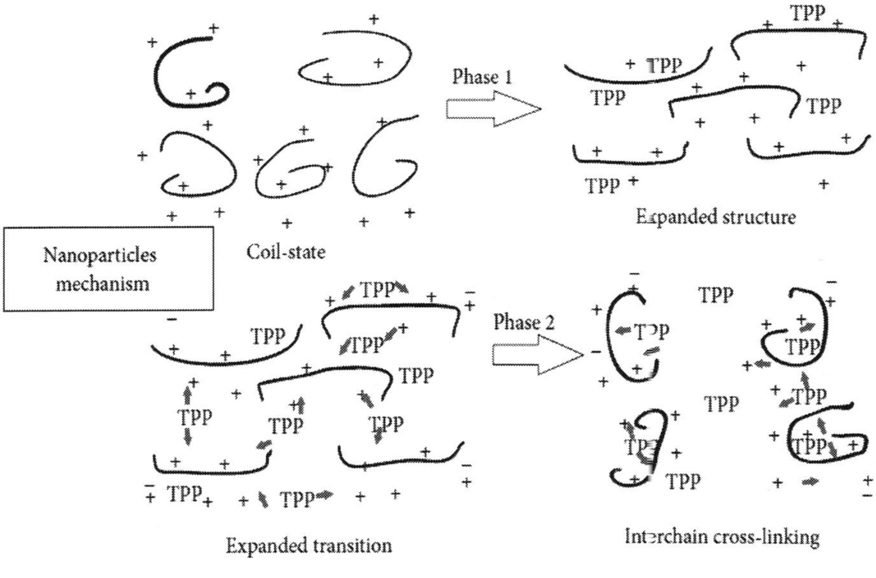

Figure 11 Two-phase mechanism of nanoparticles formation.

Bernkop-Schn¨urch et al. [29] reported stability enhancement of nanoparticles; they prepared micro and nanoparticles of native chitosan/ tripolyphosphate. The results of ITC shows enthalpy changes at binding process (ΔH =6000 cal/mol).The cooperative model was adjusted.

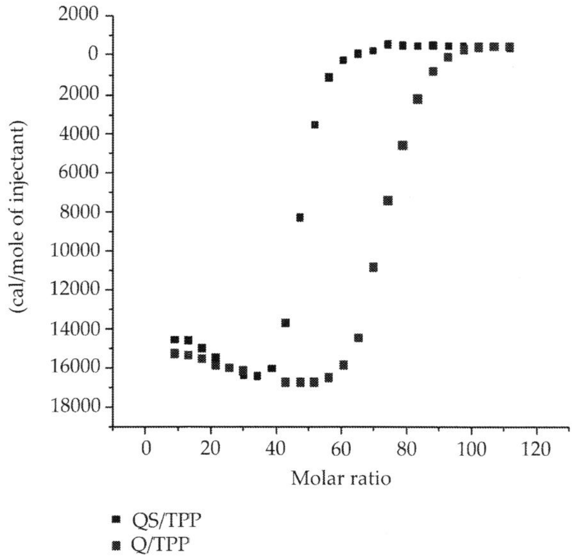

Figure 12 ITC comparative between native chitosan and thiolated chitosan.

The comparative ITC analysis between native and thiolated chitosan is shown in Figure 12. Addition of Mercapto chains reveals significant contributions of thiol groups in thermodynamic profile. The magnitude *N* [43] (Table 3) exposes the abrupt decrease of the total number of active sites; we associate this important decrease of *N* with the partial substitution in protonated amine group (NH3+). Despite important changes in the number of active sites in thiolated chitosan, the values of binding constant (Ka) were similar with native chitosan. Apparently, both reactions follow the two-phase mechanism divided at first in structural arrangement of chitosan chains, from semicoil state to expanded structure (ΔGNativeCh = −8621.78 cal∗mol ΔGThiolatedCh = −8938 cal∗mol). Gibbs free energy confirms a similar spontaneous mechanism favored to keep expanded arrangement. Additionally, in the second stage, an important increment of endothermic peak was observed after equivalent point. This energy change is due to structural reorganization of chitosan free chains, favoring semispherical shape of nanoparticles. A positive value of entropy (ΔSNativeCh = 88.7 cal∗K−1∗mol−1 ΔSThiolatedCh = 86cal∗K−1∗mol−1) suggests that water molecules are expelled from inside of chitosan and that formation of nanoparticles may happen after the saturation point [21].

CONCLUSIONS

In this work, we researched the effect of chemical derivatization on native chitosan properties by coupling reaction with 3-mercaptopropionic acid; the reaction reached 11% of modification degree of free amine groups. Chitosan derivatized with thiol groups allows preparation of stable, semispherical, and not aggregated nanoparticles, even at a high concentration of TPP. The morphologic evolution of the NPs as a function of pH solution was evaluated by AFM images, where a monodisperse population of nanoparticles and well defined semispheric structures were observed at pH 4.7. Also, a strong dependency of molar ratio (Ch-SH/TPP) with hydrodynamic size and surface charge of chitosan nanoparticles were found. In this manner, physicochemical properties of these new nanoparticles could be modulated in size and charge, between 100 nm to 200 nm, and 30mV and 40mV, respectively.On the other hand, a thermodynamic profile was carried out by ITC in order to understand the formation mechanism of nanoparticles. The reduction of magnitude (*N*) is related with the degree of substitution in amine groups. This difference could explain the variations of stability between thiolated and unmodified chitosan. We analyzed the nanoparticles formation assuming the existence of one binding site model and the aggregation process associated with conformational changes of free chitosan and thiolated chitosan chains promoted by TPP molecules. The second phase involves interchain interactions characterized by an endothermic mechanism. Additionally, according to the thermodynamic profile (Gibbs free energy and binding constant) the final cross-linking process is favored to achieve semispherical shapes nanoparticles.

REFERENCES

1. X. Wang, C. J. Summers, and Z. L. Wang, "Large-scale hexagonal-patterned growth of aligned ZnO nanorods for nano-optoelectronics and nanosensor arrays," *Nano Letters*, vol. 4, no. 3, pp. 423–426, 2004.
2. Q. Zhang and K. T. Chuang, "Adsorption of organic pollutants

from effluents of a Kraft pulp mill on activated carbon and polymer resin," *Advances in Environmental Research*, vol. 5, no. 3, pp. 251–258, 2001.

3. F. Lai, S. A. Wissing, R. H. M¨uller, and A. M. Fadda, "*Artemisia arborescens* L essential oil loaded solid lipid nanoparticles for potential agricultural application: preparation and characterization," *AAPS PharmSciTech*, vol. 7, no. 1, pp. E10–E18, 2006.
4. H. Ragelle, R. Riva, G. Vandermeulen et al., "Chitosan nanoparticles for siRNA delivery: optimizing formulation to increase stability and efficiency," *Journal of Controlled Release*, vol. 176, no. 1, pp. 54–63, 2014.
5. S. Sundar, J. Kundu, and S. C. Kundu, "Topical review: biopolymeric nanoparticles," *Science and Technology of AdvancedMaterials*, vol. 11, no. 1, Article ID014104, 13 pages, 2010.
6. J. P. Rao and K. E. Geckeler, "Polymer nanoparticles: preparation techniques and size-control parameters," *Progress in Polymer Science*, vol. 36, no. 7, pp. 887–913, 2011.
7. A Kumari, S. K. Yadav, and S. C. Yadav, "Biodegradable polymeric nanoparticles based drug delivery systems," *Colloids and Surfaces B: Biointerfaces*, vol. 75, no. 1, pp. 1–18, 2010.
8. S. K. Nitta and K. Numata, "Biopolymer-based nanoparticles for drug/gene delivery and tissue engineering," *International Journal of Molecular Sciences*, vol. 14, no. 1, pp. 1629–1654, 2013. L. Cismaru and M. Popa, "Polymeric nanoparticles with biomedical applications," *Revue Roumaine de Chimie*, vol. 55, no. 8, pp. 433–442, 2010.
9. J. L. Perry, K. G. Reuter, M. P. Kai et al., "PEGylated PRINT nanoparticles: the impact of PEG density on protein binding, macrophage association, biodistribution, and pharmacokinetics," *Nano Letters*, vol. 12, no. 10, pp. 5304–5310, 2012.
10. A Rampino, M. Borgogna, P. Blasi, B. Bellich, and A. Ces`aro, "Chitosan nanoparticles: preparation, size evolution and stability," *International Journal of Pharmaceutics*, vol. 455, no. 1 2, pp. 219–228, 2013.
11. F. Danhier, E. Ansorena, J. M. Silva, R. Coco, A. Le Breton, and V. Pr´eat, "PLGA-based nanoparticles: an overview of biomedical

applications," *Journal of Controlled Release*, vol. 161, no. 2, pp. 505–522, 2012.

12. B. Mohamed and R. Entsar, "A Biopolymer chitosan and its derivatives as promising antimicrobial agents against plant pathogens and their applications in crop protection," *International Journal of Carbohydrate Chemistry*, vol. 2011, Article ID 460381, 29 pages, 2011.
13. S. Chawla and S. Kanatt, "Chitosan," in *Polysaccharides*, pp. 1–24, Springer, 2014.
14. G. S. Dhillon, S. Kaur, S. J. Sarma, S. K. Brar, M. Verma, and R. Y. Surampalli, "Recent development in applications of important biopolymer chitosan in biomedicine, pharmaceuticals and personal care products," *Current Tissue Engineering*, vol. 2, no. 1, pp. 20–40, 2013.
15. P. Calvo, C. Remu˜n´an-L´opez, J. L. Vila-Jato, andM. J. Alonso "Novel hydrophilic chitosan polyethylene oxide nanoparticles as protein carriers," *Journal of Applied Polymer Science*, vol. 63, no. 1, pp. 125–132, 1997.
16. W. Fan, W. Yan, Z. Xu, and H. Ni, "Formation mechanism of monodisperse, low molecular weight chitosan nanoparticles by ionic gelation technique," *Colloids and Surfaces B: Biointerfaces*, vol. 90, no. 1, pp. 21–27, 2012.
17. P. K. Dutta, J. Duta, and V. S. Tripathi, "Chitin and chitosan: chemistry, properties and applications," *Journal of Scientific and Industrial Research*, vol. 63, no. 1, pp. 20–31, 2004.
18. E. Robles, E. Villar, M. Alatorre-Meda et al., "Effects of the hydrophobization on chitosan–insulin nanoparticles obtained by an alkylation reaction on chitosan," *Journal of Applied Polymer Science*, vol. 129, no. 2, pp. 822–834, 2013.
19. V. K. Mourya and N. N. Inamdar, "Chitosan-modifications and applications: opportunities galore," *Reactive & Functional Polymers*, vol. 68, no. 6, pp. 1013–1051, 2008.
20. R. Emmanuel, J. Josu´e, T. Pablo, M. V´ıctor, and V. Miguel, "Properties of insulin–chitosan complexes obtained by an alkylation reaction on chitosan," *Journal of Applied Polymer Science*, vol. 131, no. 6, 2014.

21. N. M. Alves and J. F. Mano, "Chitosan derivatives obtained by chemical modifications for biomedical and environmental applications," *International Journal of Biological Macromolecules*, vol. 43, no. 5, pp. 401–414, 2008.
22. H. Sashiwa and S.-I. Aiba, "Chemically modified chitin and chitosan as biomaterials," *Progress in Polymer Science*, vol. 29, no. 9, pp. 887–908, 2004.
23. D.-W. Lee, S. A. Shirley,R.F.Lockey, andS.S.Mohapatra, "Thiolated chitosan nanoparticles enhance anti-inflammatory effects of intranasally delivered theophylline," *Respiratory Research*, vol. 7, article 112, 2006.
24. G.Millotti, C. Samberger, E. Fr¨ohlich, D. Sakloetsakun, and A. Bernkop-Schn¨urch, "Chitosan 4-mercaptobenzoic acid: synthesis and characterization of a novel thiolated chitosan," *Journal of Materials Chemistry*, vol. 20, no. 12, pp. 2432–2440, 2010.
25. A.Anitha,N.Deepa, K. P. Chennazhi, S.V.Nair,H. Tamura, and R. Jayakumar, "Development of mucoadhesive thiolated chitosan nanoparticles for biomedical applications," *Carbohydrate Polymers*, vol. 83, no. 1, pp. 66–73, 2011.
26. W. Dhadhang, F. Achmad, and A. Ratna, "Synthesis of thiolated chitosan as matrix for the preparation ofmetformin hydrochloride microparticles," *Research in Pharmacy*, vol. 2, no. 1, pp. 26–35, 2012.
27. E.-B. Ko, H.-Y. Cho, T.-H. Kim, C.-H. Yea, and J.-W. Choi, "Cell chip with a thiolated chitosan self-assembled monolayer to detect the effects of anticancer drugs on breast normal and cancer cells," *Colloids and Surfaces B: Biointerfaces*, vol. 112, pp. 387–392, 2013.
28. A. Bernkop-Schn¨urch, M. Hornof, and D. Guggi, "Thiolated chitosans," *European Journal of Pharmaceutics and Biopharmaceutics*, vol. 57, no. 1, pp. 9–17, 2004.
29. M. Prabaharan, "Review paper: chitosan derivatives as promising materials for controlled drug delivery," *Journal of Biomaterials Applications*, vol. 23, no. 1, pp. 5–36, 2008.
30. F. Atyabi, F. Talaie, and R. Dinarvand, "Thiolated chitosan nanoparticles as an oral delivery system for amikacin: in vitro and

ex vivo evaluations," *Journal of Nanoscience and Nanotechnology*, vol. 9, no. 8, pp. 4593–4603, 2009.

31. X.Zhu,M. Su, S. Tang et al., "Synthesis of thiolated chitosan and preparation nanoparticles with sodium alginate for ocular drug delivery," *Molecular Vision*, vol. 18, pp. 1973–1982, 2012.
32. Y. Boonsongrit, B. W. Mueller, and A. Mitrevej, "Characterization of drug-chitosan interaction by 1H NMR, FTIR and isothermal titration calorimetry," *European Journal of Pharmaceuticsand Biopharmaceutics*, vol. 69, no. 1, pp. 388–395, 2008.
33. P. Yousefpour, F. Atyabi, E. Vasheghani-Farahani, A.-A. M. Movahedi, and R. Dinarvand, "Targeted delivery of doxorubicin-utilizing chitosan nanoparticles surface-functionalized with anti-Her2 trastuzumab," *International Journal of Nanomedicine*, vol. 2011, article 6, pp. 1977–1990, 2011.
34. Y. Huang and Y. Lapitsky, "Monovalent salt enhances colloidal stability during the formation of chitosan/tripolyphosphate microgels," *Langmuir*, vol. 27, no. 17, pp. 10392–10399, 2011.
35. B. Han, Y.Wei, X. Jia, J. Xu, and G. Li, "Correlation of the structure, properties, and antimicrobial activity of a soluble thiolated chitosan derivative," *Journal of Applied Polymer Science*, vol. 125, no. 2, pp. E143–E148, 2012.
36. E. S. de Alvarenga, "Characterization and properties of chitosan," in *Biotechnology of Biopolymers*, M. Elnashar, Ed., vol. 24, chapter 5, p. 364, InTech, Rijeka, Croatia, 2011.
37. L.Vijapur and S. Sreenivas, "Thiolated chitosan a gift to mucoadhesive drug delivery system—a review," *InternationalJournal of Pharmaceutical Innovations*, vol. 3, no. 4, pp. 85–95, 2013.
38. R. Martien, B. Loretz, A. M. Sandbichler, and A. B. Schn¨urch, "Thiolated chitosan nanoparticles: transfection study in the aco-2 differentiated cell culture," *Nanotechnology*, vol. 19, no. 4,Article ID 045101, 9 pages, 2008.
39. M. Clara, L. Ruo, and C. Gianluca, "Chitosan nanoparticles as therapeutic protein nanocarriers: the effect of ph on particle formation and encapsulation efficiency," *Polymer Composites*, vol. 34, no. 9, pp. 1539–1545, 2013.
40. M. A. Torres, M. M. Beppu, and C. C. Santana, "Characterization

of chemically modified chitosan microspheres as adsorbents using standard proteins (bovine serumalbumin and lysozyme)," *Brazilian Journal of Chemical Engineering*, vol. 24, no. 3, pp. 325–336, 2007.

41. X.-F. Li,X.-Q. Feng,S.Yang, T.-P.Wang, andZ.-X. Su, "Effects of molecular weight and concentration of chitosan on antifungal activity against aspergillus niger," *Iranian Polymer Journal*, vol. 17, no. 11, pp. 843–852, 2008.
42. E. Lee, J. Lee, and S. Jon, "A novel approach to oral delivery of insulin by conjugating with low molecular weight chitosan," *Bioconjugate Chemistry*, vol. 21, no. 10, pp. 1720–1723, 2010.

5
CHAPTER

PREPARATION OF CHITOSAN NANOPARTICLES BY TPP IONIC GELATION COMBINED WITH SPRAY DRYING, AND THE ANTIBACTERIAL ACTIVITY OF CHITOSAN NANOPARTICLES AND A CHITOSAN NANOPARTICLE–AMOXICILLIN COMPLEX

Tan Vui Nguyen[1], Thi Thu Ha Nguyen[1], San-Lang Wang[3], Thi Phuong Khanh Vo[1] Anh Dzung Nguyen[1]

[1]Institute of Biotechnology and EnvironmentTay Nguyen UniversityBuon Ma ThuotVietnam

[2]Department of Chemistry, Life Science Development CentreTamkang UniversityNew TaipeiTaiwan

ABSTRACT

Chitosan nanoparticles were prepared from chitosan with various molecular weights by tripolyphosphate (TPP) ionic gelation combined with a spray drying method. The morphologies and characteristics of chitosan nanoparticles were determined by TEM, FE-SEM and from their mean sizes and zeta potentials. The effect of chitosan molecular weight (130, 276, 760 and 1200 cPs) and size of spray dryer nozzle (4.0, 5.5 and 7.0 lm) on mean size, size distribution and zeta potential values of chitosan nanoparticles was investigated. The results showed that the mean size of chitosan nanoparticles was in the range of 166–1230 nm and the zeta potential value ranged from 34.9 to 59 mV, depending on the molecular weight of chitosan and size of the spray dryer nozzles. The lower the molecular weight of chitosan, the smaller the size of the chitosan nanoparticles and the higher the zeta potential. A test for the antibacterial activity of chitosan nanoparticles (only) and a chitosan nanoparticle–amoxicillin complex against Streptococcus pneumoniae was also conducted. The results indicated that a smaller chitosan nanoparticle and higher zeta potential showed higher antibacterial activity. The chitosan nanoparticle– amoxicillin complex resulted in improved antibacterial activity as compared to amoxicillin and chitosan nanopaticles alone. Using a chitosan nanoparticle–amoxicillin complex could reduce by three times the dosage of amoxicillin while still completely inhibiting S. pneumoniae.

INTRODUCTION

Streptococcus pneumoniae is a Gram-positive, normal inhabitant of the human upper respiratory tract. The bacteria can cause pneumonia, usually of the lobar type, paranasal sinusitis and otitis media which are usually secondary to one of the former infections. It also causes osteomyelitis, septic arthritis, endocarditis, cellulitis and brain abscesses. S. pneumonia currently is one of the leading causes of invasive bacterial disease in children. Particularly, S. pneumonia is strongly resistant to antibiotics. Therefore, looking for the best way to control this bacterium is a very important obligation in the pharmaceutical field [1–3].

Chitosan is a natural polymer of beta 1,4-glucosamine and beta 1,4-N-acetyl glucosamine residues. Chitosan is a non-toxic, biocompatible, biodegradable and natural polycationic agent and has great potential for agricultural [4–7] and biomedical applications, such as a promising carrier for proteins, drug delivery, vaccine delivery and antitumor treatment [8–12]. Chitosan nanoparticles are nanosized, have a large surface area, a high zeta potential, and they can provide superior activities. Recently, chitosan nanoparticles have been studied and applied in agriculture; Sang et al. [13] applied chitosan nanoparticles for coffee seedlings in the green house to enhance the growth and resistance of diseases. In the biomedical field, chitosan nanoparticles have been applied strongly as agents of drug delivery [14–17], vaccine delivery [18–20] and gene delivery [21, 22].

Bacterial resistance to antibiotics has been addressed by looking for newantibiotics and modified antibiotics. Unfortunately, the development of new antibiotics can not catch up to the microbial pathogen's fast and frequent development of resistance in a timely manner. Antimicrobial nanomaterials have been one of the recent ways to resolve this challenge, and microbial pathogens may not be able to develop the resistance and a novel nanosized form for efficient delivery. The antibacterial activity of chitosan nanoparticle–antibiotic complexes as nanoantibiotics has been recently reported. Chitosan nanoparticles prepared by an ionic gelation method with tripolyphosphate (TPP) exhibited antibacterial activity much more than a chitosan solution and doxycycline antibiotics [23]. The antibacterial activity of oleoyl–chitosan nanoparticles was also reported [24]. Anal et al. [25] also investigated the preparation of ionotropic cross-linked chitosan microspheres with TPP for controlled release of ampicillin, and the minimum inhibitory control (MIC) of the microspheres with Staphylococcus aureus were 12.5 lg/mL and prolonged for more 24 h. Preparation of chitosan microspheres was reported by Cevher et al. [26]. In this research, chitosan microspheres containing vancomycin were prepared by a spray drying method and it was found that the microspheres loaded with vacomycin were more effective than the intramuscular injection of vancomycin [26]. Jafarinejad et al. [27] developed chitosan nanoparticles for pulmonary delivery of itraconazole by using ionic gelation and a spray drying method. Ngan et al. [28] also reported preparation of chitosan nanoparticles by a spray drying method and loaded amoxicillin. The results showed that the antibacterial activity of chitosan nanoparticle–amoxicillin complex against S. aureus was two times more effective than the amoxicillin only [28].

The aims of this work are to determine the effect of chitosan molecular weight and nozzle size on average size, size distribution and zeta potential values of the chitosan nanoparticles, and to evaluate the antibacterial activity of the chitosan nanoparticles and chitosan nanoparticle–amoxicillin complex on S. pneumonia.

EXPERIMENTAL MATERIALS

Chitosan was purchased from Taehoon Bio, Seoul, Korea. The degree of chitosan deacetylation was approximately 80–90 % as determined by infrared (IR) spectroscopy [29]. The molecular weight of chitosan was determined via its viscosity (Brookfield viscometer): very high molecular weight (VHMW) 1200 cPs (centipoises); high molecular weight (HMW) 760 cPs; medium molecular weight (MMW) 276 cPs; and low molecular weight (LMW) 130 cPs. Streptococcus pneumoniae was obtained from the Institute of Hygiene and Epidemiology of Tay Nguyen University, Buon Ma Thuot, Dak Lak Province, Vietnam. TPP and acetic acid were obtained from Merck (Germany), and amoxicillin trihydrate was obtained from Purimox (India).

Preparation of chitosan nanoparticles Chitosan nanoparticles were prepared according to a modified ionic gelation method described by Gan et al. [21]. In brief, chitosan with various molecular weights (130, 276, 760 and 1200 cPs) was dissolved in 0.35 % (w/v) acetic acid

solution and kept over night at room temperature. The chitosan solution was adjusted to pH 5.5 by 0.5-M NaOH solutions. 0.25 % (w/v) TPP solution was dissolved in deionized water. And then, chitosan nanoparticles were prepared by dropping 0.25 % TPP solution into the chitosan solution at a chitosan:TPP ratio of 6:1 (w/w) while stirring at 900 rpm with a magnetic stirrer for 60 min at room temperature. The suspensions were centrifuged at 5000 rpm to remove precipitants and then the chitosan nanoparticles were collected by a nano spray dryer (Buchi Nano B90, Switzerland) placed in a clean bench room with three differently sized spray drying nozzles (4.0, 5.5 and 7.0 lm); the flow rate was 120 ml min-1; the drying gas flow was 1.3 m3 min-1; the inlet temperature was 120 _C; and the outlet temperature was 80 _C. The chitosan nanoparticles were packed in glass tubes and stored at 5 _C.

Particles Size, Zeta Potential Value and Morphology of Chitosan Nanoparticles

The morphologies of the chitosan nanoparticles were investigated by field emission scanning electron microscopy (FE-SEM; model JSM 7401F, JEOL, Ltd., Japan) at an acceleration voltage of 15 kV and transmission electron spectroscopy (TEM; JEM-1400, JEOL, Ltd., Japan). Zeta potential value and size distribution of the chitosan nanoparticles were determined by a Zetasizer Nano ZS (Malvern, UK).

Evaluation of the Antibacterial Activity of Chitosan Nanoparticles

Streptoccocus pneumoniae was cultured in a brain heart infusion broth (BHI) with or without 1.5 % agar (Difco Laboratories, Detroit, MI, USA). Streptoccocus pneumoniae were grown in liquid BHI medium (pH = 7.0) supplemented with 60 lg/mL of chitosan nanoparticles with different molecular weight: 130, 276, 760 and 1200 cPs at 37 _C under 5 % CO2 for 24 h. The growth inhibition of the bacteria by the chitosan nanoparticles was evaluated by a turbidity method at 610 nm. The experiment was conducted with five tests at 130, 276, 760, 1200 cPs and a control (the medium only) in triplicate, three tubes for each test.

Evaluation of Antibacterial Activity of Chitosan Nanoparticle–Amoxicillin Complex

The chitosan nanoparticle that performed the best in the above antibacterial activity test was used for this test. The liquid BHI medium (pH = 7.0) was supplemented with 60 lg/mL of chitosan nanoparticles loaded with amoxicillin at doses of 0, 5, 10, 15, 20 and 25 lg/mL. The bacteria were inoculated at 37 _C for 24 h under 5 % CO2. The growth of the bacteria was determined by a turbidity method at 610 nm.

The experiment was performed with three tests with chitosan nanoparticles only; amoxicillin only and the chitosan nanoparticle–amoxicillin complex were tested in triplicate, three tubes for each test.

Statistical Analysis

Statistical analysis was performed using one-way analysis of variance (ANOVA), followed by Duncan's multiple range tests (triplicate) by MSTATC software. A P value B 0.05 was considered statistically significant.

RESULTS AND DISCUSSION

The effects of the molecular weight of chitosan on the characteristics of chitosan nanoparticles prepared by TPP ionic gelation combined with a spray drying method and their resulting antibacterial activities are discussed herein.

Previous works reported that the size and zeta potential of chitosan nanoparticles and their antibacterial activities depended strongly on the molecular weight [9, 21, 28]. Therefore, in this work, four different molecular weights of chitosan were used for preparation of chitosan nanoparticles and evaluation of their antibacterial activity on S. pneumonia. The results, listed in Table 1, show that the size and zeta potential value of chitosan nanoparticles prepared by TPP ionic gelation method combined with spray drying depended clearly on the molecular weight. Increasing the molecular weight of chitosan material from 130 to 1200 cPs led to an increase in the average size of the nanoparticles from 166.7 to 1230 nm and a decrease in the zeta potential value from 54.7 to 34.9 mV. The results were better than those in previous works [21, 28, 30]. Gan et al. [21] prepared chitosan nanoparticles by TPP ionic gelation from low- , medium- and high-molecular weight chitosan. The results showed that when increasing the molecular weight of chitosan, the size of the particles increased from 136 to 1628 nm and zeta potential decreased from 48.3 to 14.3 mV [21].

The size distribution and zeta potential value of the chitosan nanoparticles are shown in Figs. 1 and 2. The results shown in Fig. 1 indicated that all peaks of size distribution in various molecular weights were sharp. It means that the size distribution of the nanoparticles was narrow and the particles were the same size and uniform. These results are also demonstrated by TEM and SEM in Figs. 5 and 6.

The results shown in Table 1 and Fig. 2 also indicated that the chitosan nanoparticles prepared by the method possesed values of zeta potentials 34.9 to 54.7 mV and higher than the chitosan nanoparticles prepared by spray drying in our previous work [28]. The results shown in Table 1 and Fig. 1 also indicated that higher molecular weight of chitosan was higher size and lower zeta potential of chitosan nanoparticles.

Table 1 Relationship between the molecular weight of chitosan and the mean size, zeta potential of the nanoparticles and their antibacterial activity

Viscosity molecular weight (cps)	Mean size of the nanoparticles (nm)	Zeta potential of the nanoparticles (mV)	OD 610 nm
130	166.7	54.7	0.078 ± 0.0069^{a}
276	189.7	47.7	0.089 ± 0.0018^{b}
760	334.6	38.6	0.115 ± 0.0060^{c}
1200	1230.0	34.9	0.127 ± 0.0146^{d}
Control[f]	–	–	0.336 ± 0.0074^{e}

The mean size of the chitosan nanoparticles were 166.7 nm, 334.6 nm and 1230.0 nm at molecular weight of chitosan of 130, 760 and 1200 cPs. In general, the results indicated that the size distribution, zeta potential value and morphology of the chitosan nanoparticles prepared by TPP ionic gelation combined with a spray drying method in this work were better than the previous works [26–28, 30–34].

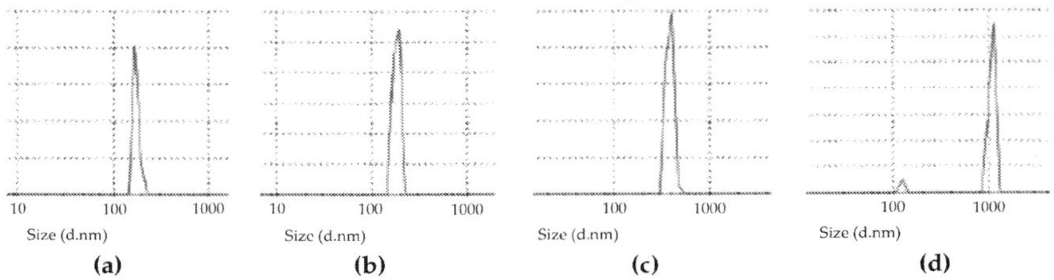

Fig. 1 Effect of various molecular weights of chitosan on the size distribution of chitosan nanoparticles prepared by TPP ionic gelation combined with a spray drying method under conditions: molecular weight of chitosan: 130 cPs (**a**); 276 cPs (**b**); 760 cPs (**c**) and 1200 cPs (**d**) at nozzle size 5.5 μm. Chitosan solution 0.1 % (w/v) in acetic acid 0.35 %, pH of the chitosan solution: 5.5 adjusted by NaOH 0.5 M; TPP (Tripolyphosphate) solution 0.25 % (w/v) in deionized water. The spray dryer was operated at a flow rate of 120 mL min^{-1}; the drying gas flow rate was 1.3 m^3 min^{-1}; the inlet temperature was 120 °C; and the outlet temperature was 80 °C

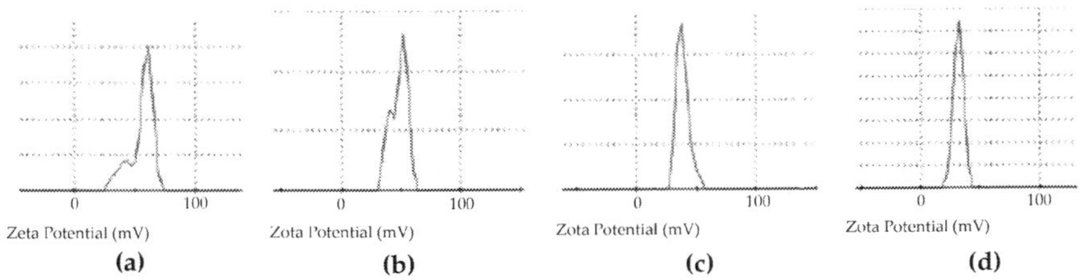

Fig. 2 Effect of various viscosity molecular weights of chitosan on zeta potential of chitosan nanoparticles prepared by TPP ionic gelation combined with a spray drying method. Molecular weight of chitosan: 130 cPs (**a**); 276 cPs (**b**); 760 cPs (**c**); 1200 cPs (**d**); and at a nozzle size of 5.5 μm

Dudhani and Kosaraju [32] prepared chitosan nanoparticles from low-molecular weight chitosan (20–200 cPs) by a TPP ionic gelation method; the weight ratio of chitosan to TPP was 2:1. Dudhani and Kosaraju reported that size distribution of the nanoparticles were very high, ranged from 20 to 600 nm and the zeta potential was only 31 mV [32]. Kasˇpar et al. [34] used the same method to prepared chitosan nanoparticles with low-molecular weight chitosan. The size distribution of the chitosan particles was so big, from 1 to 8 lm, and the particles were nonuniform and not smooth. But the zeta potential of the particles ranged only from 20 to 24 mV [34]. The antibacterial activity of the chitosan nanoparticles shown in Table 1 indicated that chitosan nanoparticles can inhibit strongly the growth of S. pneumoniae at a concentration of 60 lg/mL. The turbidity (OD 610 nm) of the test supplemented 60 ug/mL of chitosan nanoparticles with 130 cPs molecular weight was 0.078 compared to 0.336 in the control. The obtained results showed that higher molecular weight of chitosan nanoparticles and lower antibacterial activity on S. pneumoniae. Previous works reported that the antibacterial activity of chitosan depends on the polycationic charge on the chitosan molecule, higher positive charge and higher antibacterial activity [8, 9, 11]. The results shown in Table 1 demonstrated that chitosan nanoparticles prepared from low-molecular weight chitosan (130 cPs) had the highest zeta potential value (54.7 mV) and introduced highest antibacterial activity. These results were similar to some previous works [28, 35]. Fernandes et al. [35] studied the effect of the molecular weight of chitosan (625, 590 and 107 kDa) and a chitosan oligomer (1, 3 and 5 kDa) on antibacterial activity on E. coli and S. aureus; Fernandes reported that lower molecular weight chitosan and higher antibacterial activity [35]. Ngan et al. [28] prepared chitosan nanoparticles by spray drying with molecular weights of 276, 760 and 1200 cPs. The average size of particles was from 95 to 335 nm; a decrease in zeta potential values from 45.7 mV to 29.6 mV was observed, and an increase in the turbidity of the bacteria's culture (S. aureus) from 0.014 to 0.065 at 20 lg/mL was also reported.

Effect of nozzle size on characteristics of the chitosan nanoparticles and their antibacterial activity on Streptococcus pneumonia

In this experiment, chitosan nanoparticles were prepared by ionic gelation combined with a spray drying method with 130-cPs chitosan material; other conditions were the same as the above experiment, but with different nozzle sizes: 4.0, 5.5 and 7.0 lm. The results shown in Table 2 indicate that using different nozzle sizes affects the size of the chitosan nanoparticles and their zeta potential values. Although the nanoparticles were formed by TPP ionic gelation, the size of the particles may be changed in the spray drying step (Figs. 3–6). Increasing the nozzle size from 4.0 to 7.0 lm resulted in an increase in the average size of the particles from 156.1 to 376.1 nm, and a decrease in zeta potential values from 59.3 mV to 50.0 mV. However, these changes were not as high as the above experiment. These results were similar to those of Kasˇpa's work [34], in which chitosan particles were prepared by the same method with two steps: TPP ionic gelation and spray drying with 1.4- and 1.5-mm nozzle caps.

Table 2 Effect of nozzle size on characteristics of the nanoparticles prepared by ionic gelation combined with a spray drying method

Observed data	Nozzle size (μm)		
	4.0	5.5	7.0
Mean size (nm)	156.1	186.7	376.1
Zeta potential (+mV)	59.3	54.7	50.0
Antibacterial activity (OD 610 nm)	0.053 ± 0.0009	0.0625 ± 0.0010	0.0792 ± 0.0025

Chitosan nanoparticles prepared by ionic gelation combined with a spray drying method under conditions: 0.1 % chitosan solution with 130 cPs; 0.25 % TPP; the ratio of chitosan:TPP was 6:1 (w/w); chitosan nanoparticles were collected by a nano spray dryer at a flow rate of 120 mL min^{-1}; the drying gas flow rate was 1.3 m^3 min^{-1}; the inlet temperature was 120 °C and the outlet temperature was 80 °C at nozzle sizes of 4.0, 5.5 and 7.0 μm

The results showed the size of nozzle cap clearly affected the size of the particles and zeta potential values. The size distribution of the particles ranged from 0.7 to 7.0 lm with a 1.4 mm nozzle cap, and increased to 3.0–9.0 lm with a 1.5-mm nozzle cap, much bigger than our results. Zeta potential values were 20 to 24 mV, lower than those of this article [34].

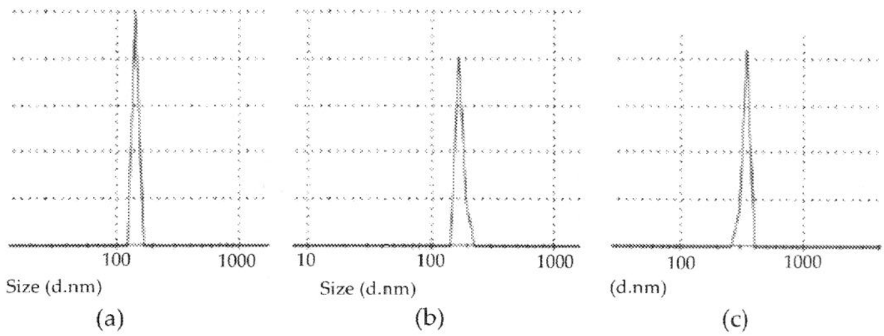

Fig. 3 Effect of nozzle size on the size distribution of chitosan nanoparticles prepared by TPP ionic gelation combined with a spray drying method. Molecular weight of chitosan: 130 cPs. Chitosan solution 0.1 % (w/v) in acetic acid 0.35 %, pH of the chitosan solution: 5.5 adjusted by NaOH 0.5 M; TPP (Tripolyphosphate) solution 0.25 % (w/v) in deionized water. The spray dryer was operated at a flow rate of 120 mL min^{-1} ; the drying gas flow rate was 1.3 m^3 min^{-1}; the inlet temperature was 120 °C; and the outlet temperature was 80 °C with various nozzle sizes: 4.0 μm (a); 5.5 μm (b) and 7.0 μm (c)

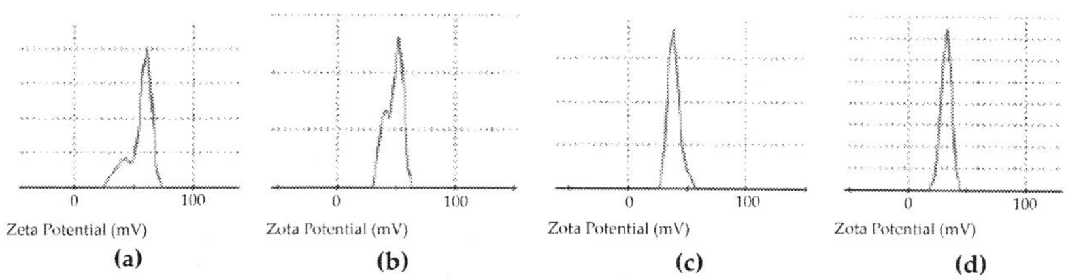

Fig. 4 Effect of nozzle size on zeta potential of chitosan nanoparticles prepared by TPP ionic gelation combined with a spray drying method under conditions: 0.1 % (w/v) chitosan solution 130 cPs was dissolved in acetic acid 0.35 %, pH of the chitosan solution was 5.5 adjusted by NaOH 0.5 M; TPP (tripolyphosphate) solution 0.25 % (w/v) in deionized water. The spray dryer was operated at a flow rate of 120 mL min^{-1}; the drying gas flow rate was 1.3 m^3 min^{-1} ; the inlet temperature was 120 °C; and the outlet temperature was 80 °C with various nozzle sizes: 4.0 μm (a); 5.5 μm (b) and 7.0 μm (c)

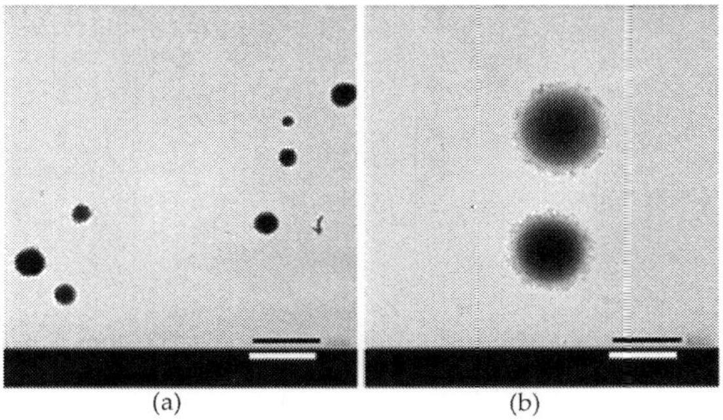

Fig. 5 TEM of of chitosan nanoparticles prepared by prepared by TPP ionic gelation combined with a spray drying method under conditions: 0.1 % chitosan solution 760 cPs (*left*) and 1200 cPs (*right*); 0.25 % TPP; the ratio of chitosan:TPP was 6:1 (w/w); chitosan nanoparticles were collected by a nano spray dryer (Buchi Nano B90, Switzerland) operated at a flow rate of 120 mL min^{-1} the drying gas flow rate was 1.3 m^3 min^{-1}; the inlet temperature was 120 °C; the and outlet temperature was 80 °C with a nozzle size of 5.5 μm

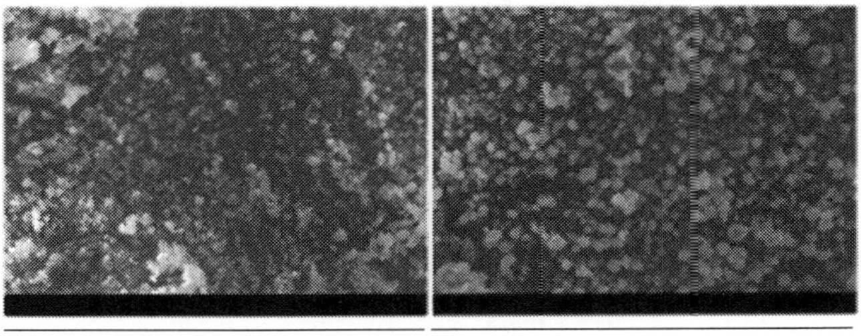

Fig. 6 SEM of chitosan nanoparticles prepared by prepared by TPP ionic gelation combined with a spray drying method under conditions: 0.1 % chitosan solution 130 cPs (*left*) and 1200 cPs (*right*); 0.25 % TPP; the ratio of chitosan:TPP was 6:1; chitosan nanoparticles were collected by a nano spray dryer operated at a flow rate of 120 mL min^{-1}; the drying gas flow rate was 1.3 m^3 min^{-1}; the inlet temperature was 120 °C; and the outlet temperature was 80 °C with a nozzle size of 7.0 μm

The antibacterial activity of the chitosan nanoparticles (Table 2) showed that the activity depends on the zeta potential value of chitosan nanoparticles. The particles had higher zeta potential values and higher growth inhibition of the bacteria (Figs. 5, 6).

Antibacterial activity of chitosan nanoparticle-amoxicillin complex

60-lg chitosan nanoparticles with 130 cPs molecular weight were mixed with 5, 10, 15, 20 and 25 lg/mL amoxicillin to make the complex. The results showed that chitosan nanoparticles with a molecular weight of 130 cPs were able to load over 90 % of the amoxicillin. The antibacterial activity of the chitosan nanoparticle– amoxicillin complex is shown in Fig. 7. The results indicate that amoxicillin can completely inhibit growth of S. pneumoniae at 25 lg/mL.

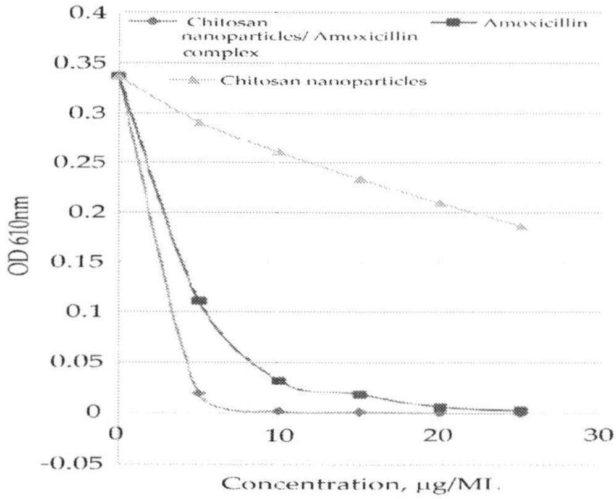

Figure 7 Effect of a chitosan nanoparticle-amoxicillin complex on growth inhibition of streptococcus pneumoniae. 130-cPs chitosan nanoparticles were loaded with amoxicillin dosages of 0,5 15, 20 and 25µg/mL and added into the cultivation medium of the bacteria. the bacteria were inoculated at 37°C for 24 h under 5% CO_2. The growth of the bacteria was determined by a turbidity method at 610 nm

The antibacterial activity of the chitosan nanoparticle–amoxicillin complex increased strongly in comparison to amoxicillin and chitosan nanoparticles alone. The bacteria could be inhibited completely at 7.5 lg/mL of amoxicillin mixed with 60-lg chitosan nanoparticles. It means the MIC value of the chitosan nanoparticle–amoxicillin complex on S. pneumonia is three times lower than the amoxicillin alone. The interactive effect of chitosan nanoparticles and antibiotics is similar to those reported by Anal et al. [25], Cevher et al. [26] and Ngan et al. [28]. The obtained results suggest that a chitosan nanoparticle–amoxicillin complex is a promising nanoantibiotic with advantages such as reduced side effects, low antimicrobial resistance, controlled drug release, low immunosuppression, improved solubility and low cost [36].

CONCLUSIONS

The size and zeta potential value of chitosan nanoparticles prepared by TPP ionic gelation combined with a spray drying method are easily controlled by manipulating the molecular weight of the chitosan material and nozzle size. The size of chitosan nanoparticles ranges from 160 to

1230 nm with a homogeneous shape, and the zeta potential values range from 35 to 59 mV. The antibacterial activity of a chitosan nanoparticle–amoxicillin complex is three times stronger than the amoxicillin alone. It is concluded that a chitosan nanoparticle–amoxicillin complex is a promising nanoantibiotic.

REFERENCES

1. R. Dagan, Vaccine 19(Suppl 1), S9–S16 (2000).
2. J. Nishant, S. Meenu, T.K. Kumar, B. Bhavneet, A. Amit, K. Ajay, K. Harpreet, C. Neelima, PLoS One 9, e96282 (2014).
3. T. van der Poll, S.M. Opal, Lancet 374, 1543–1556 (2009).
4. A.T. Rodriguez, M.A. Ramirez, R.M. Cardenas, A.N. Hernandez, M.G. Velazquez, S. Bautista, Pestic. Biochem. Physiol. 89, 206–215 (2007).
5. L.Q. Luan, V.T.T. Ha, N. Nagasawa, T. Kume, F. Yoshii, T.M. Nakanishi, Biotechnol. Appl. Biochem. 41, 49–57 (2005).
6. N.A. Dzung, V.T.P. Khanh, T.T. Dung, Carbhydr. Polym. 84, 751–755 (2011).
7. T. Yamaguchi, Y. Maehara, O. Kodama, M. Okada, M. Matsumura, N. Shibuya, J. Plant Physiol. 159, 1147–1149 (2002).
8. M. Rinaudo, Prog. Polym. Sci. 31, 603–632 (2006).
9. H.K. No, N.J. Park, S.H. Lee, S.P. Meyers, Int. J. Food Microbiol. 74, 65–72 (2002).
10. M.N. Alves, J.F. Mano, Int. J. Biol. Macromol. 43, 401–414 (2008).
11. D.N. Ngo, S.K. Kim, in Antioxidant, antimicrobial properties of chitin, chitosan, and their derivatives, ed. by S.K. Kim. Chitin and Chitosan Derivatives: Advances in Drug and Discovery and Developments (CRC Press, 2014), pp. 201–209.
12. N.A. Dzung, in Chitosan and Chitosan Derivatives as potential adjuvants for Influenza Vaccine, ed. by S.K. Kim. Chitin and Chitosan Derivatives, Advances in Drug and Discovery and Developments, in Advances in Drug and Discovery and Developments (CRC Press, 2014), pp. 418–492.
13. N.V. Sang, D.M. Hiep, N.A. Dzung, Biocatal. Agric. Biotechnol. 2,

289–294 (2013).

14. M.R. Avadi, A.M.M. Sadeghi, A. Tahzibi, K. Bayati, M. Pouladzadeh, M. Zohuriaan, M. Rafiee- Tehrani, Eur. Polym. J. 40, 1355–1361 (2004).
15. A.J. Kevin, M.P. Fresneau, A. Marazuela, A. Fabra, M.J. Alonso, J. Control. Release 73, 255–267 (2001).
16. S.A. Agnihotri, N. Mallikarjuna, T.M. Amineabhavi, J. Control Release 100, 5–28 (2004).
17. H. Li, L. Qin, Z. Wang, S. Li, Res. Chem. Intermed. 38, 1421–1429 (2012).
18. N.K. Gupta, P. Tomar, V. Sharma, V.K. Dixit, Vaccine 29, 9026–9037 (2011).
19. B. Guy, N. Pascal, A. Francon, Vaccine 19, 1794–1805 (2001).
20. O. Borges, J. Tavares, A. de Sousa, G. Borchard, H.E. Junginger, A. Cordeiro-da-Silva, 2007. Eur. J. Pharm. Sci. 32, 278–290 (2007).
21. Q.T. Gan, C. Wang, P.M. Cochrane, Colloids Surf. B 44, 65–73 (2005).
22. T. Kiang, J. Wen, H.W. Lim, K.W. Leong, Biomaterials 25, 5293–5301 (2004).
23. L. Qi, Z. Xu, X. Jiang, C. Hu, X. Zou, Carbohydr. Res. 339, 2693–2700 (2004).
24. K. Xing, X.G. Chen, Y.Y. Li, C.S. Liu, C.G. Liu, D.S. Cha, H.J. Park, Carbohydr. Polym. 74, 114–120 (2008).
25. A.K. Anal, W.F. Stevens, C. Remunan-Lopez, Int. J. Pharm. 312, 166–173 (2006).
26. E. Cevher, Z. Orhan, L. Mulazimoglu, D. Sensoy, M. Alper, A. Yildiz, Y. Ozsoy, Int. J. Pharm. 317, 127–135 (2006).
27. S. Jafarinejad, K. Gilani, E. Moazeni, M. Ghazi-Khansani, A.R. Najafabadi, N. Mohajel, Power Technol. 222, 65–70 (2012).
28. L.T.K. Ngan, S.L. Wang, D.M. Hiep, P.M. Luong, N.T. Vui, T.M. Dinh, N.A. Dzung, Res. Chem. Intermed. 40, 2165–2175 (2014).
29. G.A. Robert, J.G. Domzy, Makromol. Chem. 186, 1671–1677 (1985).
30. T.P. Learoyd, J.L. Burrows, E. French, P.C. Seville, Eur. J. Pharm.

Biopharm. 68, 224–234 (2008).

31. B.N. Estevinho, F. Rocha, L. Santos, A. Alves, Trends Food Sci. Tech. 31, 138–155 (2013).
32. A.R. Dudhani, S.L. Kosaraju, Carbohydr. Polym. 81, 243–251 (2010).
33. P. de Vos, M.M. Farr, M. Spasojevic, J. Sikkema, Int. Dairy J. 20, 292–302 (2010).
34. O. Kasˇpar, M. Jakubec, F. Stepanek, Power Technol. 240, 31–43 (2013).
35. J.C. Fernandes, F.K. Tavaria, J.C. Soares, O.S. Ramos, M.J. Monteiro, M.E. Pintado, F.X. Macata, Food Microbiol. 25, 922–928 (2008).
36. A.J. Huh, Y.J. Kwon, J. Control. Release 156, 128–145 (2011).

6

CHAPTER

UNDERSTANDING THE MECHANISM OF IONIC GELATION FOR SYNTHESIS OF CHITOSAN NANOPARTICLES USING QUALITATIVE TECHNIQUES

Sijumon Kunjachan[1,2], Sajan Jose[1], Twan Lammers[2]

[1]Department of Pharmaceutical Sciences, Mahatma Gandhi University, Kerala, India,

[2]Department of Experimental Molecular Imaging (EXMI), University Clinic, RWTH Aachen, Germany

ABSTRACT

We focused on qualitatively exploring the basic mechanisms involved in the *Ionic gelation* (IG) process, a method quite frequently used for synthesis of chitosan (CS) microparticles (MPs) and nanoparticles (NPs). We synthesized CS MPs and NPs using the Ionic gelation and microemulsion methods, and characterized the CS NPs and MPs at different stages of formulation using scanning electron microscopy (SEM) and fluorescence microscopy. Fourier Transform Infrared (FTIR) analysis was carried out to confirm effective cross-linking. Moreover, for the first time, we reported the mechanisms of IG technique for CS NP and MP synthesis with qualitative proof: (1) Complex formation of long chain oligomers with polyanions (long beaded structures) (2) cleavages at weak sites on addition of acid (HCl) (3) formation of CS NPs on chain scission. The versatility of IG for the synthesis of CS MPs and NPs was proved and compared with the microemulsion technique, thereby enhancing the wide spectrum of its use in therapeutics and biomedical applications.

INTRODUCTION

Ionic gelation (IG), earlier known as '*ion-induced gelation*', results in nanoparticles (NPs) and microparticles (MPs) with defects such as improper surface morphology, fragile particulate system, high dispersibility index, and lack of proper surface modification sites to attach functional moieties.[1-3] Here, we attempt to overcome the defects associated with IG synthesis and formulate chitosan (CS) NPs, and in principle, compare them with the already established microemulsion formulation protocols and then to discern their usefulness in drug delivery applications such as pulmonary inhalations, mucoadhesive systems, oral delivery systems for treating peptic ulcers, and finally for intra-tumoral drug targeting. All above-mentioned 'defects' can be resolved by the novel insights gained from the understanding of the synthesis mechanisms[4,5] of the *IG method.* CS, a natural linear biopolyaminosaccharide obtained by alkaline deacetylation of chitin, and second abundant polysaccharide next to cellulose.[6] Rendering amino (-NH2) and hydroxyl (-OH) groups, CS enables a high degree of chemical modification. The N-deacetylated product of chitin is an interesting biopolymer to prepare MPs and NPs, owing to its unique polymeric cationic character, good biocompatibility, nontoxicity, biodegradability, its mucoadhesivity, and absorption-enhancing effects. Chitin is a straight chain homopolymer composed of (1, 4)-linked N-acetyl glucosamine units, while CS comprises of copolymers of glucosamine and N-acetyl glucosamine.[7,8] CS has one primary amino group and two free hydroxyl groups for each C6 building unit [Figure 1]. Due to the availability of free amino groups, it carries a positive charge and reacts with many negatively charged surfaces such asthe cell membrane, mucus lining (due to negatively charged sialic acid residues), and also with other anionic polymers.[9] CS is a weak base, insoluble in water and organic solvents, however, it is soluble in dilute aqueous acidic solution ($pH < 6.5$), which can convert the glucosamine units into a soluble form of protonated amine ($R–NH_3+$).[10] CS gets precipitated in alkaline solutions or with polyanions and forms a gel at a lower pH.[11] Particle size, density, viscosity, degree of deacetylation, and molecular weight are important parameters of CS, influencing its pharmaceutical formulations. The mucoadhesive properties of CS [12,13], due to molecular attractive forces formed by electrostatic interaction between positively charged CS and negatively charged mucosal surfaces is well documented.

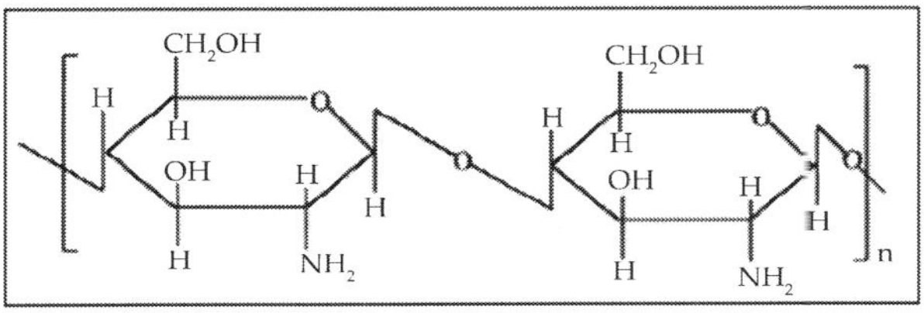

Figure 1: Schematic chemical structure of CS

These properties may be attributed to (a) strong hydrogen bonding groups like –OH, –COOH,[14] (b) strong charges,[15] (c) high molecular weight,[16] (d) sufficient chain flexibility,[17] and (e) surface energy properties favoring spreading into the mucus. CS MPs or NPs are the most widely studied drug delivery systems for controlled release of drugs, namely, antibiotics, antihypertensive agents, anticancer agents, proteins, peptide drugs, and vaccines.

MATERIALS AND METHODS

Light liquid Paraffin, heavy liquid Paraffin, hexane, methanol, sodium nitrate, and glacial acetic acid (GAA) were procured from CDH Chemicals Laboratory, India. CS (degree of deacetylation: 85% and viscosity: 800 – 1000 cps), span 85, tween 20, sodium chloride, potassium dihydrogen phosphate, and dextran sulfate were obtained from Sigma Chemicals, USA. Glutaraldehyde, toluene, orthophosphoric acid and triethylamine (TEA) were purchased from Qualigens Fine Chemicals, India. The dialysis bag membrane (MWCO: 3500) was purchased from Sigma, USA.

Optimized Procedure for the Formulation of Cs Mps Using the Microemulsion Cross-Linking Method

CS was procured at an average molecular weight ranging between 3800 and 2,000,000 Da and deacetylated to 66 to 95%, and was used for these studies. The method used for preparation was a modified protocol reported by Wang Y *et al.*[18] Light and heavy liquid paraffin oil (1:1) was taken and a mixture of Span 85, Tween 20 (7 – 8% v/v) in a ratio of 10:1 was added to this, which was then homogenized at 16,000 rpm for two minutes, to ideally disperse the surfactants in the oil base. During homogenization, the CS solution (1% w/v) was slowly injected into the oil phase to form an o/w emulsion.

Homogenization was carried out at 16,000 rpm for about 30 minutes that resulted in the formation of primary emulsion. The cross-linking agent, toluene-saturated-glutaraldehyde (1:8), was added to it stage by stage (every hour for a period of three hours). The preparation was transferred to a magnetic stirrer and left for four to five hours, for stabilization and cross-linking. The whole process was performed under light protective conditions and the temperature was maintained at 37±1°C throughout the experiment. The formulation was centrifuged at 8000 g and the sediment recovered and washed thrice with n-hexane and acetone. The residue was washed with triple distilled water. It was then freeze-dried using liquid N2 to generate rocky ice, followed by sublimation at -85°C and 05 m torr. The freeze-dried product was a fine and free flowing powder, which was collected and stored in an air-tight container at 4°C. Optimization of the procedure was carried out by varying the process parameters such as surfactant concentration, stabilizing time (using crosslinking agent), concentration of CS (aqueous phase), and homogenizing time and speed.

Optimization of Process Parameters

Several parameters such as change in surfactant concentration, change in stabilizing time, change in concentration of CS, change in homogenization time and homogenization speed were tested and optimized. Surfactant concentration was varied from 2 to 7% v/v and it was noticed that the optimized surfactant concentration (7 – 8% v/v) produced an ideal emulsion with improved dispensability. However, it was observed that a surfactant concentration of more than 8% (v/v) resulted in an agglomeration of particles and ultimately bigger particles (unpublished data). The stabilizing / cross-linking time was varied and toluene-saturated-glutaraldehyde (8:1) was added step by step. In spite of starting with two hours of cross linking time, the optimized procedure ended in four hours, during which, bound water molecules from CS MPs evaporated and cross-linking took place effectively (reaction between amino group of CS and aldehyde group of glutaraldehyde occurred eventually). Stabilizing for a period of four to five hours was found to be effective for formulating good cross-linked particles (unpublished data). Change in the concentration of the CS solution had a direct effect on the particle size of the formulation.

Different formulations using CS solution of varying concentrations were prepared and particle size was analyzed. An ideal concentration of 1% (v/v) was used for the formulation of CS MPs. Homogenization was carried out for different time intervals, and the optimized time for homogenization was 30 minutes, above which it could generate heat, produced due to shear force, and result in agglomeration of the emulsion (as observed under the light microscope). A significant change in particle size was observed with a change in homogenizing speed. Varying homogenizing speeds from 13,000 to 16,000 rpm resulted in the formulation of finer particles (unpublished data).

Cross-linked and non-cross-linked MPs

Cross-linking of MPs during formulation was carried out using toluene-saturated glutaraldehyde. Cross-linking of particles provided compact fine particles, which dispersed quite easily once formulated. During cross-linking of CS MPs, the amino group of CS reacted with the aldehyde group of glutaraldehyde and resulted in the formation of the amino-aldehyde complex (Schiffs base). Cross-linked and non-cross linked MPs were analyzed using FTIR (Qualitative approach).

Optimized Procedure for the Formulation of Cs Mps Using the Ionic Gelation Method

Degradation of high molecular weight CS into low molecular weight CS (LMWCS)

Low molecular weight oligomers were obtained by unspecific digestion. One percent (w/v) sodium nitrite solution was used to digest the CS polymer into oligomers and monomers, and incubated for one week in a shaking water bath (37°C). Digested CS was suspended in a dialysis bag membrane containing de-ionized water, which restricted the flow of oligomers, but allowed the perfusion of water soluble monomers in a water bath, thus retaining uniform-sized oligomers for the IG method, for synthesis of CS NPs.

Formulation of CS NPs using the IG method

LMWCS (1% v/v) dissolved in acetic acid solution (1% w/v) was used throughout the studies in the IG formulation method. LMWCS (5 ml) was gently dripped into the 10% v/v tripolyphosphate (TPP) and the solution was allowed to stand still for five minutes to allow gelation to occur. pH was monitored and observed to shift toward alkaline conditions (≥7.5) on addition of TPP; and was adjusted to acidic pH using 1M HCl. The precipitated NPs were washed thrice with water to remove excess unreacted TPP and freeze-dried. Mannitol 1% v/v (cryoprotectant) was used for lyophilization at -80°C using a freeze dryer.

Characterization of CS MPs and NPs

Fourier Transform Infrared spectrometry was used to analyze the structural purity of the formed CS MPs. Both cross-linked and non-cross-linked MPs were analyzed using FTIR.

RESULTS AND DISCUSSION

Chitosan microparticles were synthesized using the emulsion cross-linking technique and the particles exhibited ideal shape morphology [Figure 2].

Characterization of cross-linked and non-cross-linked MPs using

FTIR spectrum: FTIR analysis was carried out to find out the cross-linking of MPs. In emulsion cross-linking, the amino group of CS reacted with the aldehyde group of glutaraldehyde to form an amino-aldehyde complex, characterized by FTIR spectroscopy [Figures 3 and 4]. The FTIR spectrum highlighted the –NH = CH- prominence in the data.

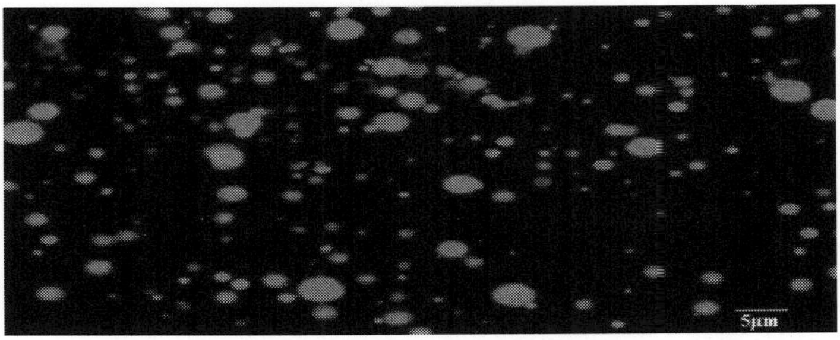

Figure 2: CS MPs prepared using the emulsion cross-linking method (intrinsic auto fluorescence of Doxorubicin)

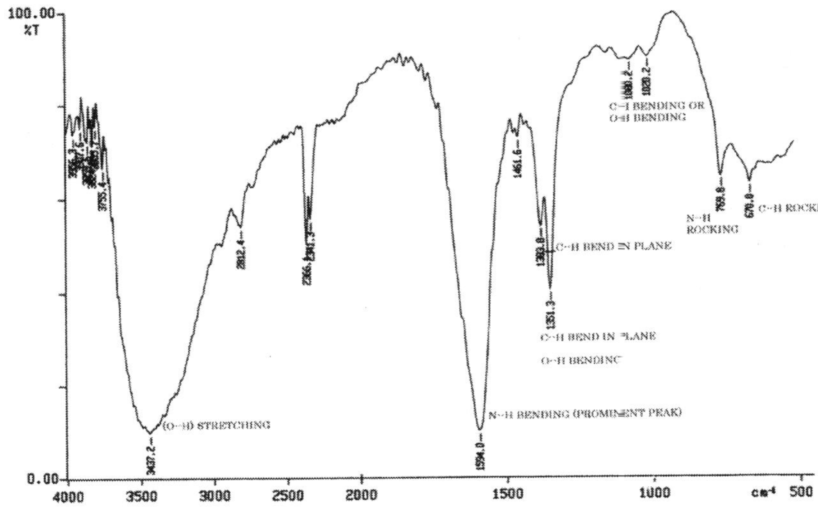

Figure 3: FTIR spectrum of CS MPs

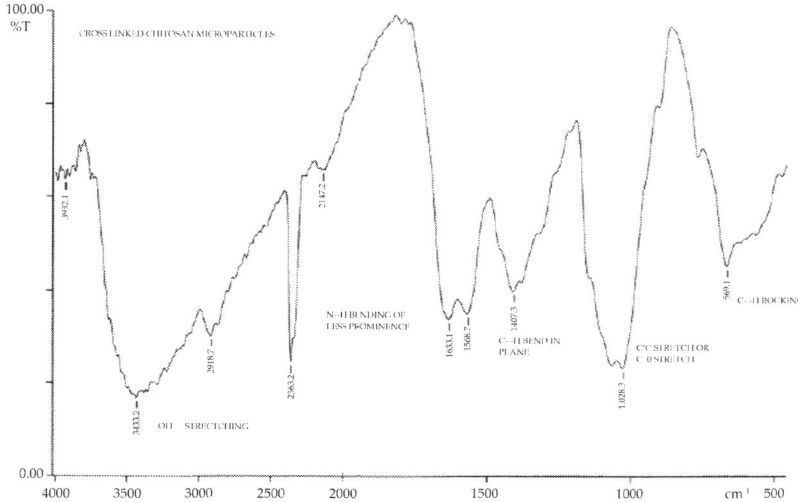

Figure 4: FTIR spectrum of cross-linked CS MPs

Hiding of the amino and the aldehyde peaks when compared with that of the non-crosslinked MPs provided an adequate proof for cross-linking of MPs. Prominent bands in the FTIR spectra of non-crosslinked CS MPs are reported in [Tables 1 and 2]. With the data procured, it was found that the (N-H) bending band and the (O-H) stretching bands were prominent. This further confirmed the presence of active –OH and –NH2 groups in the CS MPs formulated. Active hiding of the amino peak in the cross-linked CS MPs gave the evidence of cross-linking in the MPs. It was observed that (N-H) bending in this FTIR spectrum was less prominent (there was active hiding of the –NH2 group) proving the involvement of a reaction between the amino group of CS and the -CHO group of glutaraldehyde, resulting in the formation of an aldehyde-amino complex (or Schiffs base). Hence, these data clearly interpret that our optimized formulation protocol rightly generated CS MPs.

Table 1: Change in particle size with concentration of chitosan

Concentration of chitosan (%w/v)	D(0.1)	D(0.5)	D(0.9)*
0.1	0.066	0.128	0.674
0.2	0.065	0.126	0.765
0.3	0.083	0.134	1.014
1.0	0.286	0.533	1.049
2.0	0.076	0.146	1.307

*D(0.9)-Indicates the size of 90% of the particles

Table 2: Swelling Index calculations in different solvents

Solvent used	Dry weight (mg)	Wet weight (mg)	Swelling index
PBS (pH7.4)	2.040	2.230	1.01
Normal saline	2.109	2.113	1.001
Triple distilled water	2.154	2.203	1.022

In brief, CS MPs were synthesized using the IG method and the inner mechanisms were traced using our imaging approach supplemented with polymer chemistry studies. The molecular weight of CS played a crucial role in the IG method. Particle size could be controlled by varying the molecular weight of CS [Graph 1]. CS was digested to low molecular weight oligomers or monomers using the sodium nitritebased degradation technique, which ligated specifically at specific chemical links (glycosidase). These digested or degraded oligomers were further used for IG formulation studies. CS oligomers assumed a straight chain confirmation when suspended in GAA [Figure 5a], due to charge-to-charge repulsion at protonated amino groups ($-NH_3+$).

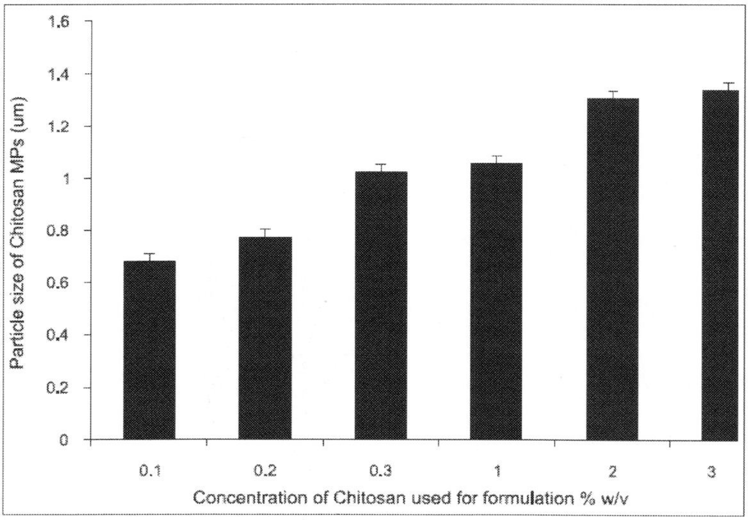

Graph 1: Change in particle size with concentration of chitosan

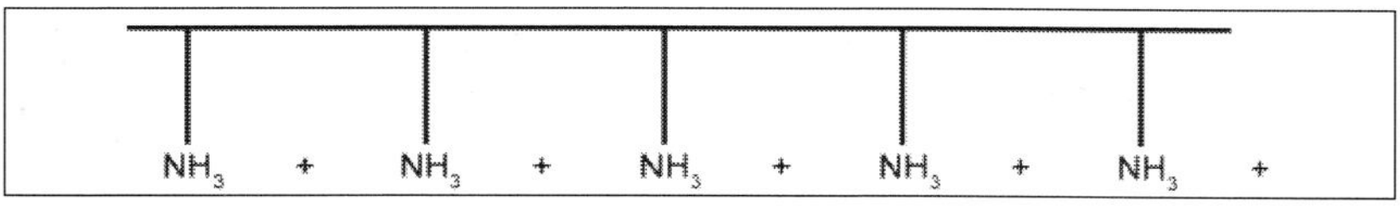

Figure 5a: CS oligomers straight chain in GAA (1%v/v)

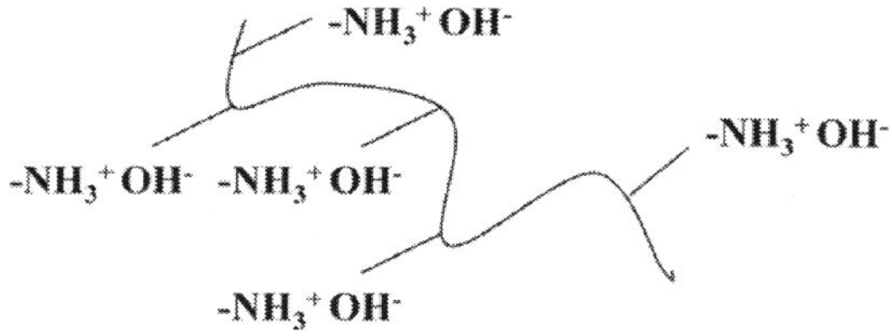

Figure 5b: Schematic diagram of CS oligomers with TPP

Formation of long chain beaded structures

On reaction with TPP, CS oligomers lose their straight chain confirmation and appear looped with or without entanglements [Figure 5b]. The degree of entanglement depends on the molecular weight of the CS used (LMWCS plays an important role). As TPP is an alkaline solution, it contains hydroxyl ions and tripolyphosphoric ions (OH-, P3O10^{5-}). There is competition between both these ions to bind to the $-NH_3+$ sites of CS.

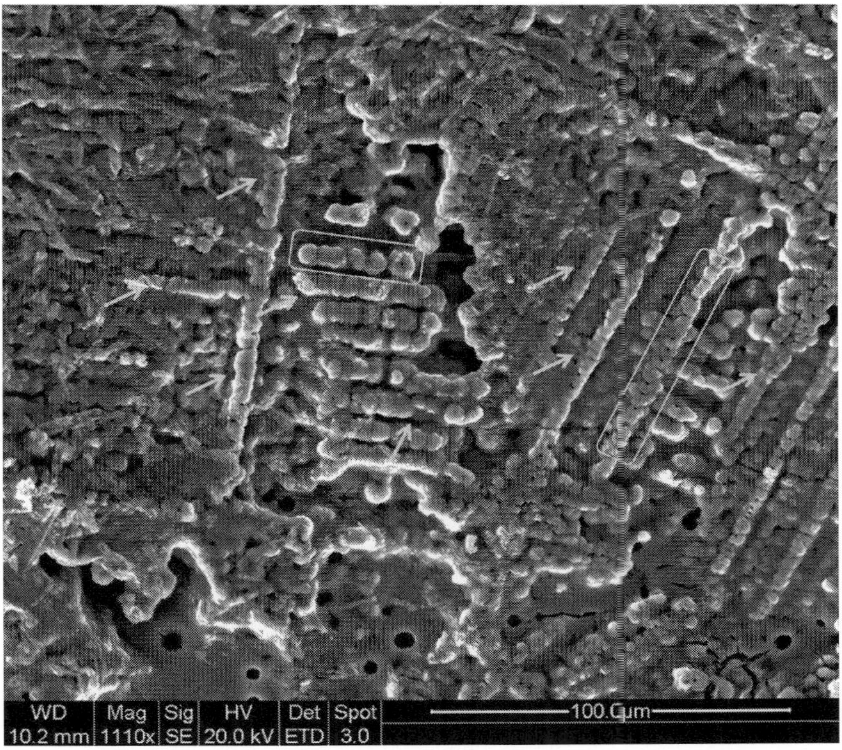

Figure 6a: SEM images of long chain CS beads

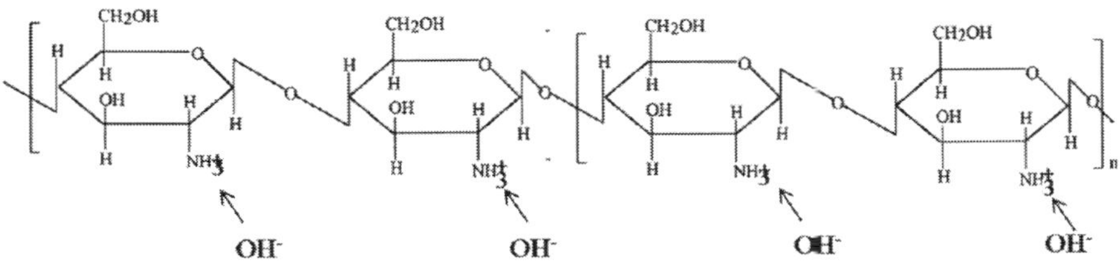

Figure 6b: Schematic chemical structure of CS with respect to long chain bead formation

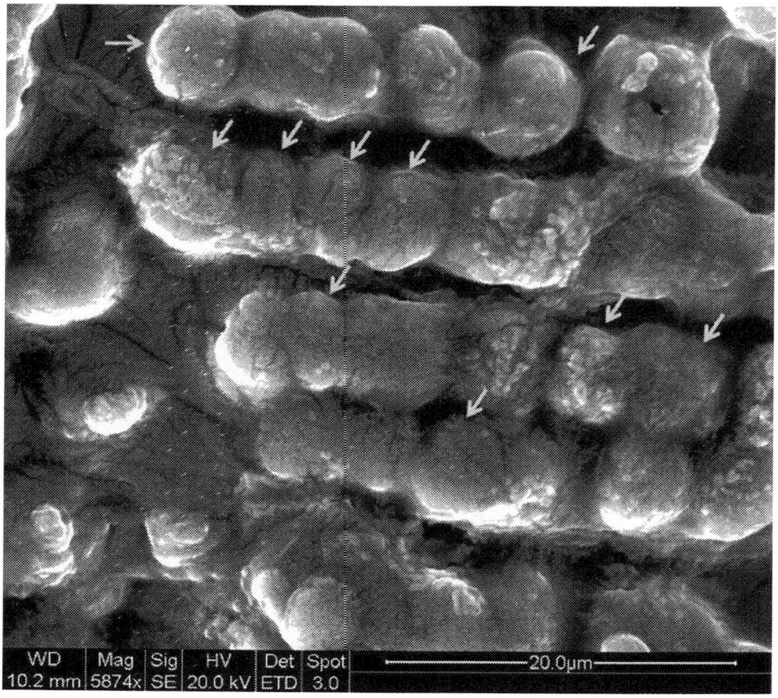

Figure 7a: SEM image showing cleaves at the weak sites of CS beaded chains (represented by arrow marks)

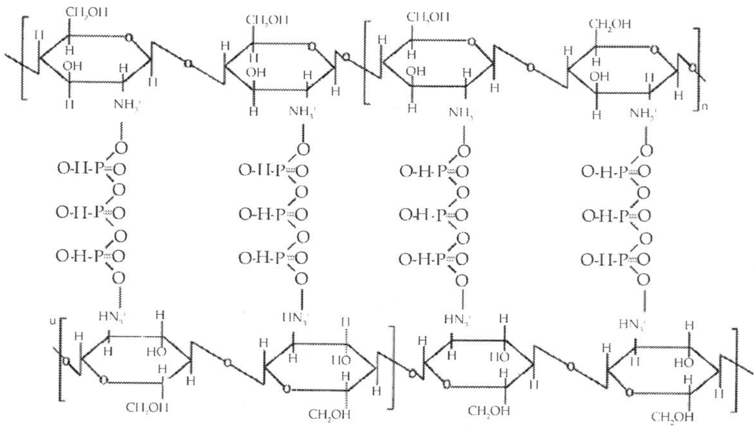

Figure 7b: Schematic representation of ionic cross linking

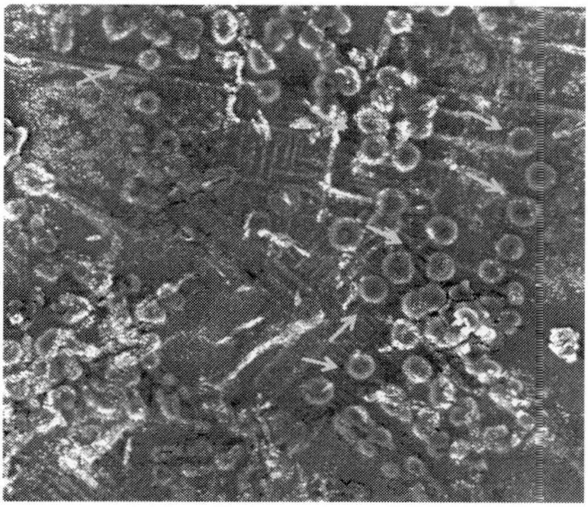

Figure 8: Formation of CS NPs (shown by marked arrows)

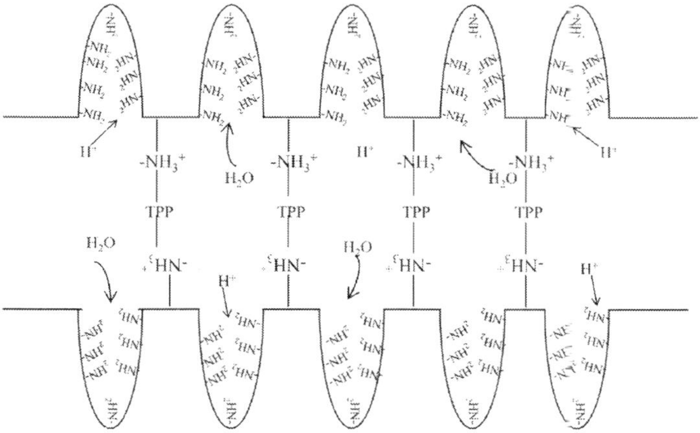

Figure 9: Schematic depiction of lower cross-linked CS bead-acid resistive property

Random attachment of both OH- and P3O10^{5-} occurs resulting in structures as shown in Figures 6a and 6b. On immediate addition of TPP, charge neutralization occurs, randomly substituting the positive charges of $-NH_3+$ with negative charges of polyphosphate ion, resulting in a complex formation of chain-like beads. They appear like a beaded chain, linked to one another, with possible sites for scission [Figure 6a and 6b]. They represent the formulation reacting with TPP at two minutes of addition of TPP. This process of chain formation is totally random, but maintains uniformity in terms of its surface morphology. The concentration of LMWCS and TPP plays a major role in this formulation step.

Formation of cleave sites

On further observation, at four minutes time interval, several points of chain scission develop on acidification [Figure 7a and 7b]. These weak sites are capable of detachment or scission (mechanical stirring or gentle shaking is also beneficial). To retain the ionic cross-linked structure and to exclude the remaining hydroxyl group, HCl is added to the solution resulting in H+ ions, which first react with $P3O10^{5-}$ ions causing "*Chain Scission*" to occur, resulting in the formation of CS MPs or NPs [Figure 8]. On further addition of HCl, certain $–NH_2$ groups exist due to the removal of water and the chain assumes a randomly coiled arrangement in the form of particles [Figure 9]. Amino ($-NH_2+$) sites act as buffer sites against protons and aid in acid resistance by formation of protonated amines ($–NH_3+$). Such systems benefit in drug targeting in gastrointestinal diseases, where drug intervention is needed at acidic pH.

Comparison of microemulsion and IG methods

Both methods exhibited *controlled cross-linking* (cross-linking with TPP and cross-linking with glutaraldehyde) depending upon the concentration of cross linkers used. The swelling index is an important parameter for iv formulations, and it also relies largely on the degree of cross-linking of the CS. Cross-linking also decides the remnant unreacted groups left for the 'acid resistive' function of CS at the acidic pH.

Particle formation is largely influenced by the pH of the TPP solution and the ionic cross-linking density. Using both the IG and microemulsion techniques, the degree of cross-linking can be modulated using simple formulation techniques such as varying the ratio of complex formation by changing the amount of TPP and CS used in case of IG, and changing the amount of toluene-saturated GAA in case of the cross-linking technique. Ideally, these formulation techniques resist the pH change (by acting as a buffer) and can be potentially used for fabricating a tumor-specific drug delivery system.

On account of its swelling property, tissue fluid can diffuse inside CS

MPs and NPs via surface pores and sense the core (which might contain the drug molecule), thereby acting as a smart polymeric drug delivery system. Although the surface morphology of CS MPs prepared by emulsion cross-linking appears fine, the size distribution is very broad and does not comply with the pharmaceutical standards. Also, polydispersity of CS MPs is far from optimum according to pharmaceutical standards per se, when compared to the IG method [Figures 2 and 8]. Apart from the simplicity of the IG method, the microemulsion formulation technique involves the use of harsh chemicals and oil bases required for the formation of primary emulsion and crosslinking, followed by its purification. However, IG uses only chitosan and TPP majorly and the excess TPP is washed off using solvents (TDW). Unlike the microemulsion method, another advantage of the IG method is the ease with which the degree of cross-linking and therefore the 'buffering' activity can be manipulated. Systematic stability studies for particles generated by both techniques are essential before it can be directly extrapolated for industrial scale up.

REFERENCES

1. Patel VM, Prajapati BG, Patel MM. Design and characterization of chitosan-containing mucoadhesive buccal patches of propranolol hydrochloride. Acta Pharm 2007;57:61-72.
2. Badawi AA, El-Laithy HM, El Qidra RK, El Mofty H, El dally M. Chitosan based nanocarriers for indomethacin ocular delivery. Arch Pharm Res 2008;31:1040-9.
3. Wu P, He X, Wang K, Tan W, He C, Zheng M. A novel methotrexate delivery system based on chitosan-methotrexate covalently conjugated nanoparticles. J Biomed Nanotechnol 2009;5:557-64.
4. Ghosn B, Kasturi SP, Roy K. Enhancing polysaccharide-mediated delivery of nucleic acids through functionalization with secondary and tertiary amines. Curr Top Med Chem 2008;8:331-40.
5. Muzzarelli RA. Chitins and chitosans as immunoadjuvants and non-allergenic drug carriers. Mar Drugs 2010;8:292-312.
6. Baxter A, Dillon M, Taylor KD, Roberts GA. Improved method for i.r. determination of the degree of N-acetylation of chitosan. Int J Biol Macromol 1992;14:166-9.

7. Singla AK, Chawla M. Chitosan: Some pharmaceutical and biological aspects–an update. J Pharm Pharmacol 2001;53:1047-67.
8. Kato Y, Onishi H, Machida Y. Application of chitin and chitosan derivatives in the pharmaceutical field. Curr Pharm Biotechnol 2003;4:303-9.
9. Onsoyen E, Skaugrud O. Metal recovery using chitosan. J Chem Technol Biotechnol 1990;49:395-404.
10. Chandy T, Sharma CP. Chitosan–as a biomaterial. Biomater Artif Cells Artif Organs 1990;18:1-24.
11. Kas HS. Chitosan: Properties, preparations and application to microparticulate systems. J Microencapsul 1997;14:689-711.
12. Jain AK, Chalasani KB, Khar RK, Ahmed FJ, Diwan PV. Muco-adhesive multivesicular liposomes as an effective carrier for transmucosal insulin delivery. J Drug Target 2007;15:417-27.
13. Needleman IG, Smales FC. *In vitro* assessment of bioadhesion for periodontal and buccal drug delivery. Biomaterials 1995;16:617-24.
14. Schipper NG, Olsson S, Hoogstraate JA, deBoer AG, Vårum KM, Artursson P. Chitosans as absorption enhancers for poorly absorbable drugs 2: Mechanism of absorption enhancement. Pharm Res 1997;14:923-9.
15. Lee YM, Kim SS, Kim SH. Synthesis and properties of poly(ethylene glycol) macromer/beta-chitosan hydrogels. J Mater Sci Mater Med 1997;8:537-41.
16. Kotzé AF, Luessen HL, de Leeuw BJ, de Boer AG, Verhoef JC, Junginger HE. Comparison of the effect of different chitosan salts and N-trimethyl chitosan chloride on the permeability of intestinal epithelial cells (Caco-2). J Control Release 1998;51:35-46.
17. Chen RH, Tsaih ML. Effect of temperature on the intrinsic viscosity and conformation of chitosans in dilute HCl solution. Int J Biol Macromol 1998;23:135-41.
18. Wang YM, Sato H, Adachi I, Horikoshi I. Optimization of the formulation design of chitosan microspheres containing cisplatin. J Pharm Sci 1996;85:1204-10.

7

CHAPTER

IMPACT OF CHITOSAN COMPOSITES AND CHITOSAN NANOPARTICLE COMPOSITES ON VARIOUS DRUG DELIVERY SYSTEMS

M.Abd Elgadir[a], Md.Salim Uddin[b], Sahena Ferdous[c], Aishah Adam, Ahmed Jalal Khan Chowdhury[c], Md.Zaidul Islam Sarker[b],

[a]Department of Pharmacology and Chemistry, Faculty of Pharmacy, Universiti Teknologi MARA, 42300 Bandar Puncak Alam, Selangor, Malaysia

[b]Department of Pharmaceutical Technology, Faculty of Pharmacy, International Islamic University Malaysia, Kuantan Campus, 25200 Kuantan, Pahang, Malaysia

[c]Faculty of Science, International Islamic University Malaysia, Kuantan Campus, 25200 Kuantan, Pahang, Malaysia

ABSTRACT

Chitosan is a promising biopolymer for drug delivery systems. Because of its beneficial properties, chitosan is widely used in biomedical and pharmaceutical fields. In this review, we summarize the physicochemical and drug delivery properties of chitosan, selected studies on utilization of chitosan and chitosan-based nanoparticle composites in various drug delivery systems, and selected studies on the application of chitosan films in both drug delivery and wound healing. Chitosan is considered the most important polysaccharide for various drug delivery purposes because of its cationic character and primary amino groups, which are responsible for its many properties such as mucoadhesion, controlled drug release, transfection, in situ gelation, and efflux pump inhibitory properties and permeation enhancement. This review can enhance our understanding of drug delivery systems particularly in cases where chitosan drug-loaded nanoparticles are applied.

INTRODUCTION

Chitosan is a natural polysaccharide and is considered the largest biomaterial after cellulose in terms of utilization and distribution [1]. It is produced from chitindthe structural element found in the exoskeleton of crustaceans such as shrimps, lobsters, and crabs. The shells of these crustaceans are first removed and then ground into powder, which is further processed to produce chitosan. Chitosan also occurs naturally in some microorganisms such as fungi and yeast [2].

Although chitosan is structurally similar to cellulose, it contains, in addition to hydroxyl groups, acetylamine or free amino groups, which display very different properties from those of cellulose [3]. Chitosan has attracted attention because of its biological properties and effective uses in the medical field, food industries, and agricultural sector [4]. It shows a variety of biological activities such as phytoalexin elicitor activity, activation of immune response, cholesterol lowering activity, and antihypertension activity [5, 6]. Similarly, mesoporous silica nanoparticles (NPs) have the ability to efficiently entrap cargo molecules because of their unique characteristic of having a huge pore size. They have already been recognized as a promising drug carrier and have recently become a new area of interest in the field of biomedical applications [7]. For instance, Zhu et al [7] focused on the stimuli-responsive controlled-release systems that responded to tumor cell environmental changes, such as pH, glucose, adenosine-50- triphosphate, glutathione, and H_2O_2.

Chitosan's therapeutic properties have also been reported by other researchers, such as inhibition of growth of microorganisms and pain alleviation [8, 9] and promotion of hemostasis and epidermal cell growth [10]. However, some researchers are interested in the potential applications of chitosan for medical and pharmaceutical purposes. The increased interest in chitosan, particularly its use in the pharmaceutical field, is attributed to its favorable properties such as biocompatibility, ability to bind some organic compounds, susceptibility to enzymatic hydrolysis, and intrinsic physiological activity combined with nontoxicity and heavy metal ions [11-13]. These properties are particularly amenable to a wide variety of biomedical applications in drug delivery and targeting, wound healing, and tissue engineering, as well as in the area of nanobiotechnology. Chitosan has attracted attention as a material for drug delivery biomedical applications in the past few years because of its biological and physicochemical properties, leading to the recognition of chitosan as a drug delivery element and a promising material specifically for the delivery of macromolecules [14-16].

In this regard, chitosan-based delivery systems range from microparticles to NP composites and films. However, there are several drawbacks in the use of chitosan for drug delivery systems. The main drawback is its poor solubility at physiological pH owing to the partial protonation of the amino groups, thereby causing presystemic metabolism of drugs in intestinal and gastric fluids in the presence of proteolytic enzymes. To overcome these inherent drawbacks, various derivatives of chitosan such as carboxylated, different conjugates, thiolated, and acylated chitosan have been used in drug delivery systems [17, 18]. Researchers reported on the goals of using chitosan as an excipient for drug delivery systems [19-23]. Therefore, the main objective of this review is to highlight and investigate the application of chitosan and chitosan-based NP composites in drug delivery systems and to provide some insight for its future potential.

Preparation and Physicochemical Properties of Chitosan

Fig. 1 shows the structures of chitin, cellulose, and chitosan. Chitosan is recognized as a linear binary heteropolysaccharide composed of b-1,4-linked glucosamine with various degrees of N-acetylation of glucosamine residues [24, 25]. It is prepared from chitin by alkaline N-deacetylation [24, 26] using concentrated sodium hydroxide (NaOH) solutions at high temperatures for a long period. Another method for the production of chitosan is N-deacetylation using enzymes under relatively mild conditions [27].

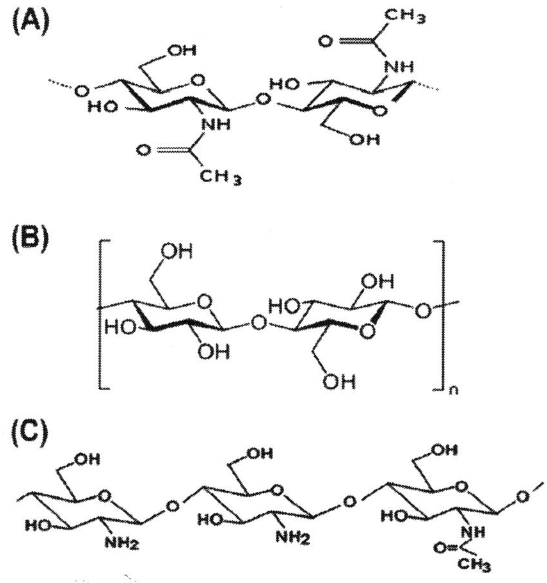

Figure 1 Structures of (A) chitin, (B) cellulose, and (C) Chitosan.

The commercially available chitosan is mostly derived from chitin of crustaceans by alkaline N-deacetylation because it is easily obtainable [28]. The production of chitosan involves a two-step process. The first step is extraction of chitin [a linear chain consisting of Nacetyl-D-glucosamine (2-acetamido-2-deoxy-b-D-gluconopyranose) joined together by b (1/4) linkage] and removal of calcium carbonate ($CaCO_3$) from crustaceans' shells using dilute hydrochloric acid and deproteination with dilute aqueous NaOH. In the second step, 40-50% aqueous NaOH at 110-115^{0}C is used for deacetylation of chitin for several hours without oxygen. When the degree of deacetylation exceeds 50%, then chitosan is produced [29]. Chitin with a degree of deacetylation of 75% is also recognized as chitosan [28]. The degree of deacetylation and molecular weight are the two fundamental parameters that can affect the properties and functionality of chitosan [26, 30]. These properties include solubility, viscosity, reactivity of proteinaceous material coagulation, and heavy metal ion chelation [31-33], and physical properties of films formulated using chitosan such as tensile strength, elasticity, elongation, and moisture absorption [34]. Chitosan is soluble in aqueous acidic solutions, but insoluble in both water and alkaline solutions [25]. The majority of polysaccharides are usually found neutral or negatively charged in an acidic environment. When dissolved, the amino groups (-NH_2) of the glucosamine are protonated to -NH_3 group [35], and the cationic polyelectrolyte readily forms electrostatic interactions with other anionic groups [36].

Therefore, the cationic chitosan molecule interacts with negatively charged surfaces that modify its physicochemical characteristics [2, 37]. These modifications of chitosan molecules are the source of its unique functional properties.

DRUG DELIVERY PROPERTIES OF CHITOSAN

Anionic Drug Delivery Properties

When a technique of drug discharge cannot be achieved by using a simple drug dissolution process such as diffusion, membrane layer handle along with erosion as well as osmotic systems, retardation mediated by ionic relationships is often used. The latter technique can be carried out with regard to cationic drugs by using anionic polymeric excipients such as polyacrylates, alginate, or carboxymethylcellulose salt. However, in anionic drug delivery systems, chitosan is the sole selection. Chitosan was used as a medication provider matrix to investigate medication release devices for the anionic medication naproxen [38]. It was found that the interactions between chitosan and the therapeutic agent was more evident, and stable complexes can also be formed from which this medicine can be produced, actually spanning a more extended period counted on an ionic cross-linking. For example, the delivery systems of enoxaparin/ chitosan nanoparticulate provided more stable complexes and resulted in significantly improved drug uptake [39]. Some anionic polymeric excipients such as carrageenan, pectin, alginate, and polyacrylates can be homogenized with chitosan, leading to high-density, relatively stable complexes. However, a similar result can be achieved by homogenizing chitosan with an alternative to multivalent anionic and inorganic polymer anions such as sulfate or tripolyphosphate (TPP) [40].

Mucoadhesive properties

The mucoadhesive properties of chitosan are probably attributable to its cationic character. Furthermore, hydrophobic interactions may help with the mucoadhesive components. The mucoadhesive properties of chitosan are weak as compared with various anionic polymeric excipients such as hyaluronic acid, polycarbophil, and carbomer [41]. In order to attain substantial mucoadhesive attributes, a polymer should have high cohesive properties because adhesive bond normally fails within the mucoadhesive polymer as opposed to involving the polymer along with the mucus gel layer. Regarding chitosans, these cohesive properties tend to be comparatively weak. It may be improved by the formation of complexes with multivalent anionic drug treatments, multivalent anionic polymeric excipients, and also multivalent inorganic anions. This strategy is effective to a very limited extent, as the cationic substructures of chitosan being accountable for mucoadhesion via ionic interactions while using the mucus are blocked in such cases.

Table 1. Selected studies on utilization of chitosan composites for drug delivery applications.

Name of chitosan composites	Purpose of utilization	Findings
Chitosan-based responsive hybrid nanogels	Integration of optical pH-sensing	Chitosan-based responsive hybrid nanogels exhibit a nonreversible pH-sensitive property and a significant cytotoxicity after 24 h treatment. It is critical to construct highly stable biopolymer-hybrid nanogel quantum dots [43].
Chitosan–zinc–pectin composite	Delivery of resveratrol to the colon	Formulation prepared at pH 1.5, 1% chitosan, 120 min cross-linking time, and pectin/drug ratio of 3:1 demonstrated the best colon-specific drug release [79].

Chitosan oligomer–zidovudine composite	Prevents disappearance of Zidovudine in human plasma and prolong its shelf life	The study indicated longer mean retention time for chitosan oligomer–zidovudine composite with values of about 1.5 h vs. 0.59 h for zidovudine alone. Chitosan oligomer–zidovudine composite was found to have accumulated in the kidney other than heart, liver, spleen, lung, and brain, and the drug had a shelf life of 12 h [66].
Chitosan–sodium alginate tablet	Vaginal delivery of chlorhexidine digluconate	Tablet containing 6% chitosan, 24% sodium alginate gave the best result of drug release [70].
Chitosan–cyclosporin A	Management of extraocular diseases	Enhancement of therapeutic index of clinically challenging drugs with potential application at extraocular level and achievement of fast drug release and therapeutic concentrations in external ocular tissues during a period of 24 h [122].
Chitosan–polyelectrolyte films	Delivery of drug for skin	The films gave significantly different drug release and drug permeation through the skin [123].

Lueben et al [42] demonstrated a significantly improved oral bioavailability involving buserelin when being administered in rats with mucoadhesive polymers, for instance, chitosan and carbomer. However, this particular effect could not be attained anymore when chitosan was mixed with polyanionic carbomer in the same formulation. More cationic character of the polymer is provided by the trimethylation of the primary amino group of chitosan. It was found that when trimethylated chitosan is added to PEGylated, its mucoadhesive properties were improved up to 3.4-fold [43]. The mucoadhesive properties of chitosan can be substantially improved as a result of the immobilization of thiol groups on it. It was reported that chitosan is able to form disulfide bonds with mucus glycoproteins when found with the mucus gel layer, and this phenomenon makes it the most mucoadhesive polymer [44].

Gelling Properties

As hydrogels form, one advantage of in situ gelling properties can be achieved when the pH-dependent hydrostability of chitosan is properly addressed. Gupta and Vyas [45] improved an in situ gelling delivery system by using a mixture of polyacrylic acid and chitosan. They observed that the resulting formulation was in a liquid state at pH 6.0 even though the same formulation underwent a rapid transition to viscous gel phase at pH 7.4. Further improvements through thiolation may also enhance the in situ gelling characteristics of chitosan. As a result of the access of oxygen on mucosal surfaces, for instance, nasal mucosa or ocular surfaces, immediately after the mixture is applied in liquid form using oxygen-free single unit forms, a cross-linking process via disulfide bond formation takes place, causing a significant increase in viscosity. Based on the cross-linking properties, the viscosity increased 16,500-fold in a period of 20 minutes using aqueous 1% (m/v) of chitosanethioglycolic acid conjugate [46].

Gene Expression Properties

Chitosan was also modified to improve its properties for gene expression purposes. For instance, the self-branching of chitosans was used as a strategy to improve its gene transfer properties, and this can be carried out without compromising the safety profile [47]. In this respect, self-branched trisaccharide- substituted chitosans, in addition to a self-branched molecular mass of 11-71 kDa, were synthesized, characterized, and also compared in contrast to their own linear counterparts with respect to transfection efficiency.

The results revealed that self-branched chitosans could yield gene expression levels two as well as five times greater than that of Lipofectamine and Exgen, respectively. In another investigation, thiolated chitosan forming intrachain bonds of disulfide was used as a good strategy to stabilize the chitosan/plasmid NP complex, resulting in higher stability properties toward nucleases [48]. In addition, owing to the reducing conditions of the cytoplasma, the plasmid was mainly released in the target cells because the disulfidebondswerelargely cleaved there, resulting in the release of the plasmid at the target site. The transfection rate of the thiolated chitosan/plasmid NP complex was found to be five times higher compared with that of the unmodified chitosan/pDNA NP complex. Owing to the trimethylation of the remaining primary amino groups, this strategy was further improved by raising the cationic character of thiolated chitosan [49]. Furthermore, chitosan/cyclodextrin and PEGylated chitosan NPs were identified as promising tools for DNA-based drug delivery [50, 51].

In contrast to small molecules, where a controlled release of anionic drugs can be achieved, stable complexes with chitosan can be formed using comparatively large polyanionic molecules such as small interfering RNA and DNA-based drugs. If the ratio of the cationic polymer is sufficiently high in the complex, NPs exhibiting a positive zeta potential can be formed. Because of the small size of these particles and the net positive charge, endocytosis was achieved particularly when the sizes of the particles were smaller than 100 nm [52]. From a toxicology viewpoint, chitosan is comparatively recognized as a less toxic polymer than other cationic polymers such as polyarginine, polylysine, and polyethyleneimine [53]. This property makes chitosan a promising excipient for nonviral gene delivery systems. It was reported that the bioavailability of DNA-based drugs delivered into the body can be improved if chitosaneDNA-based drug complexes are protected to some extent toward degradation by DNAses [54].

Permeation Enhancing Properties

Based on the positive charges of chitosan, it was found that these charges are responsible for the mechanism of permeation enhancement, which can interact with the cell membrane of chitosan, resulting in a structural reorganization of tight junction-associated proteins [55]. A primary amino group that led to a more pronounced cationic character using the trimethylation strategy did not lead to further improvements of permeation enhancing properties. It was demonstrated that the permeation enhancing properties and toxicity to a large extent were attributable to the structural properties of chitosan including the degree of deacetylation and molecular mass [56]. Chitosans with high molecular mass and high degree of deacetylation exhibited a comparatively higher increase in epithelial permeability, which could be due to molecular mass and other permeation enhancing polymers such as polyacrylates [57]. Various in vivo studies can be used to confirm this permeation enhancing effect. A 2-fold improvement of the oral bioavailability of ganciclovir was demonstrated owing to the coadministration of chitosan [58].

Chitosan can be combined with other permeation enhancers because it acts in a completely different manner from these enhancers, leading to an additive or even a synergistic effect.

Using this strategy, the oral bioavailability of ganciclovir could even be improved by 4-fold, using a combination of sodium dodecyl sulfate and chitosan compared with just a 2-fold improvement with sodium dodecyl sulfate alone. Recently, it was reported that chitosan NPs exhibit only in the first segment of the duodenum a permeation enhancing effect for small peptides. The permeation enhancing effect was enlarged over the entire duodenum owing to the addition of cyclodextrin [59]. However, >30-fold further improvement in the permeation enhancing properties of chitosan on certain mucosal membranes can be achieved because of thiolation [60].

SELECTED STUDIES ON UTILIZATION OF CHITOSAN COMPOSITES FOR DRUG DELIVERY SYSTEMS

Many studies have been conducted recently using chitosan as a drug delivery biomaterial to treat diseases such as cancer [61], optical diseases [62], and colon diseases [63]. Table 1 [43, 66, 70, 79, 122, 123] shows a selection of studies on the use of chitosan composites for drug delivery applications. A systematic series of N-trimethyl chitosan chloride polymer synthesized from different chitosans based on molecular weight (low, medium, and high molecular weight) have been coformulated into a hydrogel with polyethylene glycol (PEG) and glycerophosphate and investigated for nasal drug delivery [64]. The authors found that hydrogels derived from N-trimethyl chitosan with high or medium average molecular weight exhibit relatively short solegel transition times at physiologically relevant temperatures. The same hydrogels display good water-holding capacity and strong mucoadhesive potential. They revealed that an aqueous hydrogel formulation, which was derived from N-trimethyl chitosan of medium average molecular weight, appears particularly promising because it exhibited the most favorable rheological and mucoadhesive behaviors and a solegel transition that occurs at 32.5^{0}C within 7 minutes.

Chitosan was also investigated as an injectable vehicle for drug delivery in the presence of sodium bicarbonate ($NaHCO_3$) [65]. The hydrogels of chitosan/$NaHCO_3$ system showed porous morphologies with some diversification depending on the $NaHCO_3$ concentration, which affected their erosion and drug release rate behaviors. An in vivo gelation test was performed via a dorsal subcutaneous injection of chitosan/ $NaHCO_3$ solution in adult SpragueeDawley rats. Exactly 2% (w/v) of chitosan solution without NaHCO3 was also administered as a control. Sterile solutions were prepared via UV sterilization of solid chitosan powder, 0.22 mm filtration of 1% acetate acid solution and $NaHCO_3$ solutions, and sterilized chitosan solution and chitosan/$NaHCO_3$ mixtures. An aqueous urethane solution was injected intraperitoneally to anesthetize the rats. Each injection was 0.4 mL in volume and performed subcutaneously through a syringe equipped with a G2 gauge needle. The formation of in situ gels suggested that such systems have promising applications in injectable drug delivery. The drug delivery system prepared from chitosan oligomerezidovudine composites for the in vitro release of zidovudine was investigated [66]. A conjugate study was confirmed in mice plasma and renal homogenate. The pharmacokinetics study indicated a longer mean retention time for the chitosan oligomerezidovudine conjugate with values of about 1.5 hours compared with 0.59 hour for zidovudine alone. The chitosan oligomerezidovudine conjugates were found to accumulate (aside from the heart and the liver) in the lung, spleen, brain, and kidney after their in vivo administration. The study concluded that chitosan oligomerezidovudine conjugates have the potential to be developed into a renaltargeting drug delivery system.

SELECTED STUDIES ON CHITOSAN-BASED NPS FOR DRUG DELIVERY SYSTEMS

Nowadays, it is considered that nanomedicine will lead breakthroughs for the detection, diagnosis, and treatment of cancer [67]. Chitosan NPs are a drug carrier with the advantage of slow or controlled drug release, which improves drug solubility and stability, enhances efficacy, and reduces toxicity. In vitro and in vivo studies have also shown that chitosan has antitumor effects, leading to good prospects for their application as a supplementary antitumor drug and drug carrier [68].

Chitosan-based nanostructures predominantly work on the involved chemical cross-linking within the polymer chain. Earlier chitosan/silica nanocomposites were formed using the reaction of hydroxyl groups on chitosan monomers with tetramethoxysilane. The first data presented involved chitosan nanospheres for drug delivery applications [69]. The authors used the water-in-oil (w/o) emulsion method, which was followed by glutaraldehyde cross-linking of the chitosan amino groups. They produced nanospheres loaded by 5-fluorouracil, an anticancer drug. These studies further revealed the feasibility of reproducible synthesizing stable nanosized chitosan particles, which can entrap and deliver drugs [70]. One of chitosan's properties is its ability to form gel upon contact with special polyanions, a process referred to as "ionotropic gelation," which occurs as a result of the formation of intra and inter cross-linkages within/between polymer chains mediated by the polyanions.

Based on ionotropic gelation of TPP with chitosan, chitosan NPs have been developed for drug encapsulation [71, 72]. This simple technique involves mixing of the acidic phase (pH 4-6) containing chitosan with an alkaline phase (pH 7-9) containing TPP. NPs were immediately formed based on the mixing of these two phases through intra- and intermolecular linkages created between chitosan amino groups and TPP phosphates. Insulin-loaded chitosan NPs have also been successfully prepared using a TPP solution mixed with insulin and then adding the mixture to chitosan solution under constant stirring [73]. In brief, various concentrations of chitosan and TPP were dissolved in acetic acid (pH 4) and purified water, respectively. Different volumes of the TPP solution was mixed with 4 mL of the chitosan solution through a syringe needle under magnetic stirring at room temperature, and chitosan NPs were present in the suspension. Insulin-loaded chitosan NPs were formed spontaneously upon the incorporation of the TPP aqueous solution containing insulin to the chitosan acetic acid solution. The size of chitosan NPs were 300-400 nm with a surface positive charge ranging from þ54 to þ25 mV. In this study, the ability of chitosan NPs to enhance both relative bioavailability and intestinal absorption of insulin was investigated by monitoring the glucose level of plasma in alloxan-induced diabetic Wistar male rats. Various doses of insulin-loaded chitosan NPs were orally administrated. The stable positively charged chitosan NPs showed particle sizes within the range of 250-400 nm, and an insulin association ratio of up to 80%

was used. The in vitro release investigations indicated an initial burst phase that was pH sensitive.

The intestinal absorption of insulin was enhanced by chitosan NPs to a greater extent than the aqueous solution of chitosan in vivo. It was noticed that hypoglycemia was prolonged over 15 hours after the administration of 21.1 IU/kg insulin loaded in the chitosan NPs. However, the average bioavailability relative to the subcutaneous injection of free insulin solution showed up to 14.9%. In another study, different formulations of chitosan NPs produced by the ionic gelation of TPP and chitosan were investigated [74]. Drug delivery systems prepared using low molecular weight (LMW) chitosan NPs and monodisperse using the ionic gelation technique were also investigated [75]. The results showed that LMW chitosan NPs has good compatibility with erythrocytes, and they can be easily attached to the erythrocyte membrane surface. This indicates that the erythrocyte load of LMW chitosan can be used as a potential vascular drug delivery system.

The complex coacervation technique was previously used to prepare chitosaneDNA NPs [76, 77]. The phosphate and amino groups were used in a ratio between 8 and 3, respectively, in the presence of chitosan. This particle size was optimized to 100-250 nm range using a narrow distribution. It is possible that chitosane DNA NPs could partially protect the encapsulated plasmid DNA via the degradation of nuclease. Coalescence and emulsionedroplet coalescence methods were reported by Tokumitsu et al [78]. They used the principles of both emulsion cross-linking and precipitation. With this method, instead of cross-linking in the stable droplets, precipitation is elicited by allowing the coalescence of small chitosan droplets with NaOH droplets. A stable emulsion containing an aqueous solution of chitosan combined with the drug to be loaded is stated in liquid paraffin. At that time, another stable emulsion containing aqueous chitosan mixed with NaOH is produced in a similar manner. When both emulsions are combined under high-speed stirring, droplets of each and every emulsion would randomly collide [79]. The preparation of ultrafine polymeric NPs with a narrow size distribution may be achieved using a reverse micellar medium. Such particles can be prepared using the aqueous core of the reverse micellar droplets as a nanoreactor. The size of these very narrow and monodispersed reverse micellar tiny droplets normally lies between 1 nm and 10 nm

[80], which turns them into potential and promising NPs in drug delivery investigations. Amethod to encapsulate doxorubicinedextran conjugates in chitosan NPs was used by Mitra et al [81]. In this method, an organic solvent was applied to dissolve the surfactant for preparing reverse micelles. Several studies have been done on the self-assembly of chemically modified chitosan into NPs with an eye toward delivering macromolecules [82-84]. Fractional conjugation connected with PEG at a basic pH was proven to yield self-aggregation via an amide linkage to soluble chitosan [84]. After incubation in phosphate buffer saline, these kinds of aggregates could trap insulin because electrostatic interactions were developed between the unconjugated chitosan monomers and the anionic residues of protein. Table 2 [73, 77, 79, 124] shows a selection of studies on the utilization of chitosan NP composites for drug delivery systems.

Table 2. Selected studies on utilization of chitosan NP composites for drug delivery systems.

Name of chitosan NP composites	Purpose of utilization	Findings
Chitosan nanospheres loaded by 5-fluorouracil	Delivery of 5-fluorouracil for cancer treatment.	These stable nanosized chitosan particles can entrap and deliver drugs in tumor cells [79].
Chitosan–TPP NPs loaded with insulin	Delivery of insulin for diabetics	Chitosan NPs enhanced both the relative bioavailability and intestinal absorption of insulin, resulting in lower blood glucose level in rats [73].
Chitosan–DNA NPs	Delivery of encapsulated of plasmid DNA	Chitosan–DNA NPs protect the encapsulated plasmid DNA from nuclease degradation [77].

Chitosan NPs conjugated with doxorubicin–dextran complex	Delivery of encapsulated dextran–doxorubicin conjugate	These NPs target tumor cells with good efficiency[79].
Chitosan NPs labeled with fluorescein isothiocyanate–bovine serum albumin	Delivery of fluorescein	These NPs shows potentiality for drug delivery on the epithelial cells of ocular mucosa [124].

SELECTED STUDIES ON CHITOSAN FILMS FOR DRUG DELIVERY SYSTEMS

Chitosan was also used in the preparation of films for drug delivery systems [85-87]. Films prepared using chitosan have been utilized for oral delivery of many drugs such as chlorhexidine digluconate [88], 5-fluorouracil [89], mitoxantrone [90], cytarabine [91], and paclitaxel [92]. The characteristics of chitosan including the drug delivery behavior of nanocomposite films prepared from mixtures of chitosan and organic rectorite (OREC), which is a type of layered silicate, were investigated [93]. The films of chitosan and chitosan-OREC nanocomposite were prepared with different chitosan/ OREC mass ratios (2:1, 6:1, 12:1, 20:1, 50:1) and dissolved in a 2% (w/v) aqueous acetic acid to obtain 2% (w/v) chitosan and chitosan/OREC nanocomposite films. The films exhibited the strongest antibacterial behaviors. It was observed that all films showed equivalent drug release in the initial stages, but after several hours the release became slower compared to films prepared using pure chitosan. Chitosan and gelatin solutions were mixed together to obtain two final polymeric concentrations, F1 (1% w/v) and F2 (2% w/v), and the films prepared from the mixture were investigated for drug delivery [94]. The results showed that only the film based on gelatin alone provided complete drug release owing to its dissolution.

In 30 minutes, films with an excess of chitosan showed a higher release of drugdup to 83% as compared with 48% of the drug for films containing greater amounts of gelatin. Films of chitosan were prepared for dexamethasone delivery [95]. Dexamethasone was loaded in chitosan films at a percentage of 1.5 (wt.%). Later, the films were dried in a glass

Petri dish at room temperature for 1e3 days until monolayer films were obtained. An analog procedure was performed to achieve a bilayer film formation with dexamethasone. Release tests suggest that the dexamethasoneechitosan films are potential sustained-release carriers for dexamethasone. It was also found that the release time of the films was longer than that of conventional ocular topical delivery dosage forms. Moreover, a second layer of chitosan film significantly modified the drug release profile. Therefore, the monolayer dexamethasoneechitosan film might be considered a promising ocular delivery carrier for dexamethasone in hours and bilayer dexamethasoneechitosan film in weeks. Films prepared from chitosan and PEG with ciprofloxacin hydrochloride as the model drug incorporated at different concentrations were studied [96]. PEG was used in concentrations of 2.0 wt.%, 3.5 wt.%, 5.5 wt.%, and 8.0 wt.% of total films. Ciprofloxacin hydrochloride (0.1 g and 0.3 g) was loaded in the films. From the controlled release tests, it was found that the release of ciprofloxacin hydrochloride increased with PEG and decreased with the increase in the amount of drug loaded in the film. However, the cumulative release amount of the drug increased significantly. The chitosanePEG films were also found to be sensitive to pH and ionic strength. In simulated intestinal fluid, a reduction of the ciprofloxacin hydrochloride concentration from 100% to 71% with an increase in thickness of the film from 35 mm to 85 mm was observed.

SELECTED STUDIES ON WOUND HEALING BASED ON CHITOSAN

Chitosan is used as a wound healing accelerator in many studies [97-107]. It enhances the functions of inflammatory cells such as macrophages and polymorphonuclear leukocytes, as well as the production of osteopontin and leukotriene B4, transforming growth factor b1, and platelet-derived growth factor and fibroblasts [108]. Chitosan also possesses other biological activities and affects the macrophage function that favors faster wound healing [109]. Moreover, it has histoarchitectural tissue organization and displays an aptitude to stimulate cell proliferation [110]. The biological properties, especially bacteriostatic and fungistatic properties, are useful for wound treatment [111]. Films with flexible, thin, transparent properties prepared from a composite of chitosanealginate polyelectrolyte complex caused accelerationin healing of incision wounds in the rat model compared with conventional gauze dressing. It was observed that the closure rate and appearance of polyelectrolyte complextreated wounds were comparable with Opsite1-treated wounds [112]. An application of cross-linkable chitosan hydrogel on full-thickness skin incisions made on the backs of mice significantly induced wound contraction and resulted in a substantial acceleration of wound closure and healing compared with the untreated controls [98]. In another research, an early return to normal skin color in chitosan treated areas was observed [113]. Treatment with chitosan demonstrated a substantial decrease in treatment time with minimum scar formation in various animals. The biochemistry and histology of chitosan in wound healing have also been investigated [114]. It was found that silver sulfadiazine incorporated with bilayer chitosan wound dressing exhibited tremendous oxygen permeability, water uptake capability, and controlled water vapor transmission rate. The dressing showed excellent antibacterial activity when in vitro culture was performed for 1 week [115]. Chitosan has been studied widely as a wound dressing material. Acetate bandage for wound healing dressing as a topical antimicrobial dressing in mice was investigated by Burkatovskaya et al [116]. It was found that the bandage provided important benefits by reducing the number of inflammatory cells in the wound at Day 2 and Day 4 and by healing the wound especially during the early period where its antimicrobial effect is most important.

SAFETY ASSESSMENT OF BIOMEDICAL APPLICATION OF CHITOSAN AND ITS NP COMPOSITE

Although nanotechnology is a promising technology offering great benefits in the biomedical field, current knowledge on the safety of various NPs in biomedical application is not sufficient. Generally, chitosan has been considered comparatively safe because of its biodegradable and biocompatible properties. LMW chitosan is excreted through the kidney, whereas the excessive molecular weight can be degraded into fragments suitable for renal clearance [117]. However, the use of chitosan in unmodified forms is restricted because they are water-insoluble and highly viscous and have the tendency to coagulate with proteins at high pH values [118]. Chitosan NPs exhibit toxic properties, which make chitosan NPs applicable for cancer treatment. Some studies have reported the cytotoxicity effects of chitosan NPs in vitro [119,120]. A few research studies have been performed on genotoxicity effects and skin irritation. An in vivo study also reported that chitosan NPs affected the mice's survival rate [121]. However, despite several drawbacks, chitosan is considered a promising agent for drug delivery systems.

CONCLUSION

This review summarizes the biomedical application of chitosan and chitosan-based NP composites with emphasis on drug delivery systems. Chitosan is an important and amazing material that has so many applications in various fields of drug delivery systems. It is biodegradable and biocompatible, and can be found in abundance in nature from renewable sources. Recently, nanochitosan composites have acquired a remarkable advantage over their conventional counterparts owing to the presence of a huge surface area, which gives them additional properties, particularly in terms of biomedical applications. Further study on the drug delivery properties of chitosan and its NP composites may lead to the realization of more effective drug delivery systems.

REFERENCES

1. Mincea M, Negrulescu A, Ostafe V. Preparation, modification, and applications of chitin nanowhiskers: a review. Rev Adv Mater Sci 2012;30:225e42.
2. Illum L, Jabbal-Gill I, Hinchcliffe M, et al. Chitosan as a novel nasal delivery system for vaccines. Adv Drug Deliv Rev 2001;51:81e96.
3. Hwang JK, Shin HH. Rheological properties of chitosan solutions. Korea-Aust Rheol J 2000;12:175e9.
4. Li Q, Dunn ET, Grandmaison EW. Applications and properties of chitosan. In: Goosen MFA, editor. Applications of chitin and chitosan. Lancaster: Technomic Publishing Co.; 1997. p. 3e29.
5. Nishimura K, Nishimura S, Nishi N. Immunological activity of chitin and its derivatives. Vaccine 1984;2:93e9.
6. Allan GG, Altman LC, Bensinger RE, et al. Biomedical applications of chitin and chitosan. In: Zikakis JP, editor. Chitin, chitosan and

related enzymes. New York: Academic Press; 1984. p. 119e33.

7. Zhu CL, Wang XW, Lin ZZ, et al. Cell microenvironment stimuli-responsive controlled-release delivery systems based on mesoporous silica nanoparticles. J Food Drug Anal 2014;22:18e28.
8. Badawy MEI, Rabea EI, Rogge TM, et al. Synthesis and fungicidal activity of new N,O-acyl chitosan derivatives. Biomacromolecules 2004;5:589e95.
9. Balakrishnan B, Mohanty M, Umashankar PR et al. Evaluation of an in situ forming hydrogel wound dressing based on oxidized alginate and gelatin. Biomaterials 2005;26:6335e42.
10. Howling GI, Dettmar PW, Goddard PA, et al. The effect of chitin and chitosan on the proliferation of human skin fibroblasts and keratinocytes in vitro. Biomaterials 2001;22:2959e66.
11. Li Q, Dunn ET, Grandmaison EW, et al. Applications and properties of chitosan. J Biol Compat Polym 1992;7:370e97.
12. Ravi Kumar MNV. A review of chitin and chitosan applications. React Functional Polym 2000;46:1e27.
13. Wang XH, Cui FZ, Zhang YH. Preparation and characterization of collagen/chitosan matrices as potential biomaterials. J Bioactive Compatible Polym 2003;18:453e67.
14. Patel M, Shah T, Amin A. Therapeutic opportunities in colon specific drug delivery system. Crit Rev Ther Drug Carrier Syst 2007;24:147e202.
15. George M, Abraham TE. Polyionic hydrocolloids for the intestinal delivery of protein drugs: alginate and chitosan : a review. J Control Release 2006;114:1e14.
16. Bernkop-Schnu¨ rch A, Walker G. Multifunctional matrices for oral peptide delivery. Crit Rev Ther Drug Carr Syst 2001;18:459e501.
17. Chopra S, Mahdi S, Kaur J, et al. Advances and potential applications of chitosan derivatives as mucoadhesive biomaterials in modern drug delivery. J Pharm Pharmacol 2006;58:1021e32.
18. Riva R, Ragelle H, Rieux A, et al. Chitosan and chitosan derivatives in drug delivery and tissue engineering. Adv Polym Sci 2011;244:19e44.
19. Anitha A, Maya S, Deepa N, et al. Efficient water soluble

O-carboxymethyl chitosan nanocarrier for the delivery of curcumin to cancer cells. Carbohydr Polym 2011;83:452e61.

20. Chen M, Liu Y, Yang W, et al. Preparation and characterization of self-assembled nanoparticles of 6-Ocholesterol- modified chitosan for drug delivery. Carbohydr Polym 2011;84:1244e51.
21. Ferrari PC, Souzab FM, Giorgettib L, et al. In vitro drug permeation from chitosan pellets. Carbohydr Polym 2012;87:2526e31.
22. Pawar H, Douroumis D, Boateng J. Preparation and optimization of PMAAechitosanePEG nanoparticles for oral drug delivery. Colloids Surf B Biointerfaces 2012;90:102e8.
23. Termsarasab U, Cho HJ, Kim DH, et al. Chitosan oligosaccharideearachidic acid-based nanoparticles for anti-cancer drug delivery. Int J Pharm 2013;441:373e80.
24. Kittur FS, Vishu Kumar AB, Tharanathan RN, et al. Low molecular weight chitosansd preparation by depolymerization with Aspergillus niger pectinase and characterization. Carbohydr Res 2003;338:1283e90.
25. Krajewska B. Membrane-based processes performed with use of chitin/chitosan materials. Sep Purif Technol 2005;41:305e12.
26. Berger J, Reist M, Mayer JM, et al. Structure and interactions in chitosan hydrogels formed by complexation or aggregation for biomedical applications. Eur J Pharm Biopharm 2004;57:35e52.
27. Prashanth KVH. Solid state structure of chitosan prepared under different N-deacetylation conditions. Carbohydr Polym 2002;50:27e33.
28. Cervera MF, Heinamaki J, Rasanen M, et al. Solid-state characterization of chitosans derived from lobster chitin. Carbohydr Polym 2004;58:401e8.
29. Steenkamp GC, Keizer K, Neomagus HWJP, et al. Copper (II) Removal from polluted water with alumina/chitosan composite membranes. J Membr Sci 2002;197:147e56.
30. Cho J, Heuzey MC, Begin A, et al. Viscoelastic properties of chitosan solutions: effect of concentration and ionic strength. J Food Eng 2006;74:500e15.
31. Rege PR, Block LH. Chitosan processing: influence of process

parameters during acidic and alkaline hydrolysis and effect of the processing sequence on the resultant chitosans properties. Carbohydr Res 1999;321:235e45.

32. Duarte ML, Ferreira MC, Marvao MR, et al. An optimized method to determine the degree of acetylation of chitin and chitosanbyFTIRspectroscopy. Int JBiolMacromol2002;31:1e8.
33. Gamage A, Shahidi F. Use of chitosan for the removal of metal ion contaminants and proteins from water. Food Chem 2007;104:989e96.
34. Nunthanid J, Puttipipatkhachorn S, Yamamoto K, et al. Physical properties and molecular behavior of chitosan films. Drug Dev Ind Pharm 2001;27:143e57.
35. WangB,WangK,DanW,etal.Konjacglucomannanecollagenechitosan blend films. J Biomed Eng 2006;23:102e6.
36. Fee M, Errington N, Jumel K, et al. Correlation of SEC/MALLS with ultracentrifuge and viscometric data for chitosans. Eur Biophys J 2003;32:457e64.
37. Xu YX, Kim KM, Hanna MA, et al. Chitosanestarch composite film: preparation and characterization. Ind Crops Prod 2005;21:185e92.
38. Bhise KS, Dhumal RS, Paradkar AR, et al. Effect of drying methods on swelling, erosion and drug release from chitosanenaproxen sodium complexes. AAPS Pharmscitech 2008;9:1e12.
39. Sun W, Mao S, Wang Y, et al. Bioadhesion and oral absorption of enoxaparin nanocomplexes. Int J Pharm 2010;386:275e81.
40. Shavi G, Nayak U, Reddy M, et al. Sustained release optimized formulation of anastrozole-loaded chitosan microspheres: in vitro and in vivo evaluation. Mater Sci Mater Med 2011;22:865e78.
41. Grabovac V, Guggi D, Bernkop-Schnu¨ rch A. Comparison of the mucoadhesive properties of various polymers. Adv Drug Deliv Rev 2005;57:1713e23.
42. Lueßen HL, de Leeuw BJ, Langemey¨ er MWE, et al. Mucoadhesive polymers in peroral peptide drug delivery: VI. Carbomer and chitosan improve the intestinal absorption of the peptide drug buserelin in vivo. Pharm Res 1996;13:1668e72.
43. Jintapattanakit A, Junyaprasert VB, Kissel TJ, et al. The role of mucoadhesion of trimethyl chitosan and PEGylated trimethyl chitosan

nanocomplexes in insulin uptake. Pharm Sci 2009;98:4818e30.

44. Werle M, Bernkop-Schnu¨ rch A. Thiolated chitosans: useful excipients for oral drug delivery. J Pharm Pharmacol 2008;60:273e81.
45. Gupta S, Vyas SP. Carbopol/chitosan based pH triggered in situ gelling system for ocular delivery of timolol maleate. Sci Pharm 2010;78:959e76.
46. Sakloetsakun D, Hombach J, Bernkop-Schnu¨ rch A. In situ gelling properties of chitosanethioglycolic acid conjugate in thepresenceofoxidizingagents.Biomaterials 2009;30:6151e7.
47. Malmo J, Vrum KM, Strand SP. Effect of chitosan chain architecture on gene delivery:comparison of self-branched and linear chitosans. Biomacromolecules 2011;12:721e9.
48. Martien R, Loretz B, Thaler M, et al. Chitosan-thioglycolicacid conjugate: an alternative carrier for oral nonviral gene delivery? J Biomed Mater Res A 2007;82:1e9.
49. Varkouhi AK, Verheul RJ, Schiffelers RM, et al. Gene silencing activity of siRNA polyplexes based on thiolated N,N,N-trimethylated chitosan. Bioconjugate Chem 2010;21:2339e46.
50. Teijeiro-Osorio D, Remu~n_an-L_opez C, Alonso MJ. Chitosan/cyclodextrin nanoparticles can efficiently transfect the airway epithelium in vitro. Eur J Pharm Biopharm 2009;71:257e63.
51. Malhotra M, Lane C, Tomaro-Duchesneau C, et al. A novel scheme for synthesis of PEG-grafted chitosan polymer for preparation of nanoparticles and other applications. Int J Nanomed 2011;6:485e94.
52. Mao S, Sun W, Kissel T. Chitosan-based formulations for delivery of DNA and siRNA. Adv Drug Deliv Rev 2010;62:12e27.
53. Yu H, Chen X, Lu T, et al. Poly(l-lysine)-graftechitosan copolymers: synthesis, characterization, and gene transfection effect. Biomacromolecules 2007;8:1425e35.
54. Lee D, Mohapatra SS. Chitosan nanoparticle-mediated gene transfer. New York: Humana Press; 2008.
55. Schipper NGM, Olsson S, Hoogstraate JA, et al. Chitosans as absorption enhancer for poorly absorbable drugs: 2. Mechanism of absorption enhancement. Pharm Res 1997;14:923e9.
56. Schipper NGM, Varum KM, Artursson P. Chitosans as absorption

enhancers for poorly absorbable drugs. 1. Influence of molecular weight and degree of deacetylation on drug transport across human intestinal epithelial (Caco- 2) cells. Pharm Res 1996;13:1686e92.

57. Kast CE, Bernkop-Schnurch A. Influence of the molecular mass on the permeation enhancing effect of different poly(acrylates). STP Pharm Sci 2002;6:351e6.
58. Shah P, Jogani V, Mishra P, et al. Modulation of ganciclovir intestinal absorption in presence of absorption enhancers. J Pharm Sci 2007;96:2710e22.
59. Trapani A, Lopedota A, Franco M, et al. A comparative study of chitosan and chitosan/cyclodextrin nanoparticles as potential carriers for the oral delivery of small peptides. Eur J Pharm Biopharm 2010;75:26e32.
60. Langoth N, Kahlbacher H, Sch€offmann G, et al. Thiolated chitosans: design and in vivo evaluation of a mucoadhesive buccal peptide drug delivery system. Pharm Res 2006;23:573e9.
61. Kim JH, Kim YS, Park K, et al. Self-assembled glycol chitosan nanoparticles for the sustained and prolonged delivery of antiangiogenic small peptide drugs in cancer therapy. Biomaterials 2008;29:1920e30.
62. Wu W, Shen J, Banerjee P, et al. Chitosan-based responsive hybrid nanogels for integration of optical pH-sensing, tumor cell imaging and controlled drug delivery. Biomaterials 2010;31:8371e81.
63. Saboktakin MR, Tabatabaie RM, Maharramov A, et al. Synthesis and in vitro evaluation of carboxymethyl starchechitosan nanoparticles as drug delivery system to the colon. Int J Biol Macromol 2011;48:381e5.
64. Nazar H, Fatouros DG, van der Merwe SM, et al. Thermosensitive hydrogels for nasal drug delivery: the formulation and characterisation of systems based on N-trimethyl chitosan chloride. Eur J Pharm Biopharm 2011;77:225e32.
65. Liu L, Tang X, Wang Y, et al. Smart gelation of chitosan solution in the presence of NaHCO3 for injectable drug delivery system. Int J Pharm 2011;414:6e15.
66. Liang Z, Gong T, Sun X, et al. Chitosan oligomers as drug carriers for renal delivery of zidovudine. Carbohydr Polym 2012;87:2284e90.

67. Fan Z, Fu PP, Yu H, et al. Theranostic nanomedicine for cancer detection and treatment. J Food Drug Anal 2014;22:3e17.

68. Wang JJ, Zeng ZW, Xiao RZ, et al. Recent advances of chitosan nanoparticles as drug carriers. Int J Nanomed 2011;6:765e74.

69. Ohya Y, Shiratani M, Kobayashi H, et al. Release behavior of 5-fluorouracil from chitosan-gel nano-spheres immobilizing 5-fluorouracil coated with polysaccharides and their cell specific cytotoxicity. J Macromol Sci Pure Appl Chem 1994;31:629e42.

70. Janes KA, Calvo P, Alonso MJ. Polysaccharide colloidal particles as delivery systems for macromolecules. Adv Drug Del Rev 2001;47:83e97.

71. Shirashi S, Imai T, Otagiri M. Controlled release of indomethacin by chitosanepolyelectrolyte complex: optimization and in vivo/in vitro evaluation. J Control Release 1993;25:217e25.

72. Gan Q, Wang T, Cochrane C, et al. Modulation of surface charge, particle size and morphological properties of chitosan-TPP nanoparticles intended for gene delivery. Colloids Surf B Biointerfaces 2005;44:65e73.

73. Pan Y, Li Y, Zhao H, et al. Bioadhesive polysaccharide in protein delivery system: chitosan nanoparticles improve the intestinal absorption of insulin in vivo. Int J Pharm 2002;249:139e47.

74. Xu Y, Du Y. Effect of molecular structure of chitosan on protein delivery properties of chitosan nanoparticles. Int J Pharm 2003;250:215e26.

75. Fan W, Yan W, Xu Z, et al. Formation mechanism of monodisperse, low molecular weight chitosan nanoparticles by ionic gelation technique. Colloids Surf B Biointerfaces 2012;90:21e7.

76. Dodane V, Vilivalam D. Pharmaceutical application of chitosan. Pharm Sci Technol Today 1998;1:246e53.

77. Ray K, Mao HQ, Lin KY, et al. Oral immunization with DNA chitosan nanoparticles. Proc Int Symp Control Release Mater 1999;26:348e9.

78. Tokumitsu H, Ichikawa H, Fukumori Y. Chitosanegadopentetic acid complex nanoparticles for gadolinium neutron-capture therapy of cancer: preparation by novel emulsionedroplet coalescence

technique and characterization. Pharm Res 1999;16:1830e5.

79. Hamidi M, Azadi A, Rafiei P. Hydrogel nanoparticles in drug delivery. Adv Drug Deliv Rev 2008;60:1638e49.

80. Maitra A. Determination of size parameters of water aerosol OT-oil reverse micelles from their nuclear magnetic resonance data. J Phys Chem 1984;88:5122e5.

81. Mitra S, Gaur U, Ghosh PC, et al. Tumour targeted delivery of encapsulated dextranedoxorubicin conjugate using chitosan nanoparticles as carrier. J Control Release 2001;74:317e23.

82. Yu S, Hu J, Pan X, et al. Stable and pH-sensitive nanogels prepared by self-assembly of chitosan and ovalbumin. Langmuir 2006;22:2754e9.

83. Ichikawa S, Iwamoto S, Watanabe J. Formation of biocompatible nanoparticles by self-assembly of enzymatic hydrolysates of chitosan and carboxymethyl cellulose. Biosci Biotechnol Biochem 2005;69:1637e42.

84. Ohya Y, Cai R, Nishizawa H, et al. Preparation of PEG-grafted chitosan nano-particle for peptide drug carrier. Proc Int Symp Control Release Bioact Mater 1999;26:655e6.

85. Macleod GS, Collett JH, Fell JT. The potential use of mixed films of pectin, chitosan and HPMC for bimodal drug release. J Control Release 1999;58:303e10.

86. Shu XZ, Zhu KJ. The influence of multivalent phosphate structure on the properties of ionically cross-linked chitosan films for controlled drug release. Eur J Pharm Biopharm 2002;54:235e43.

87. Perugini P, Genta I, Conti B, et al. Periodontal delivery of ipriflavone: new chitosan/PLGA film delivery system for a lipophilic drug. Int J Pharm 2003;252:1e9.

88. Senel S, Ikinci G, Kas S, et al. Chitosan films and hydrogels of cholhexidine gluconate for oral mucosal delivery. Int J Pharm 2000;193:197e203.

89. Ouchi T, Banba T, Fujimoto M, et al. Synthesis and antitumor activity of chitosan carrying 5-fluorouracil Makromol Chem Physics 1989;190:1817e25.

90. Jameela SR, Jayakrisnan A. Glutaraldehyde cross-linked chitosan

microspheres as a long acting biodegradable drug delivery vehicle: studies on the in vitro release of mitoxantrone and in vivo degradation of microspheres in rat muscle. Biomaterials 1995;16:769e75.

91. Blanco MD, Gomez C, Olmo R, et al. Chitosan microspheres in PLG films as devices for cytarabine release. Int J Pharm 2000;202:29e39.

92. Miwa A, Ishibe A, Nakano M, et al. Development of novel chitosan derivatives as micellar carriers of taxol. Pharm Res 1998;15:1844e50.

93. Wang X, Du Y, Luo J, et al. Chitosan/organic rectorite nanocomposite films: structure, characteristic and drug delivery behaviour. Carbohydr Polym 2007;69:41e9.

94. Abruzzo A, Bigucci F, Cerchiara T, et al. Mucoadhesive chitosan/gelatin films for buccal delivery of propranolol hydrochloride. Carbohydr Polym 2012;87:581e8.

95. Rodrigues LB, Leite HF, Yoshida MI, et al. In vitro release and characterization of chitosan films as dexamethasone carrier. Int J Pharm 2009;368:1e6.

96. Wang Q, Dong Z, Du Y, et al. Controlled release of ciprofloxacin hydrochloride from chitosan/polyethylene glycol blend films. Carbohydr Polym 2007;69:336e43.

97. Ueno H, Mori T, Fujinaga T, et al. Topical formulations and wound healing applications of chitosan. Adv Drug Deliv Rev 2001;52:105e15.

98. Ishihara M, Nakanishi K, Ono K, et al. Photocrosslinkable chitosan as a dressing for wound occlusion and accelerator in healing process. Biomaterials 2002;23:833e40.

99. Kweon DK, Song SB, Park YY, et al. Preparation of watersoluble chitosan/heparin complex and its application as wound healing accelerator. Biomaterials 2003;24:1595e601.

100. Alemdaro_glu C, De_gimZ, C¸ elebi N, et al. An investigation on burn wound healing in rats with chitosan gel formulation containing epidermal growth factor. Burns 2006;32:319e27.

101. Minagawa T, Okamura Y, Shigemasa Y, et al. Effects of molecular weight and deacetylation degree of chitin/chitosan on wound healing. Carbohydr Polym 2007;67:640e4.

102. Hong HJ, Jin SE, Park JS, et al. Accelerated wound healing by

smad3 antisense oligonucleotides-impregnated chitosan/ alginate polyelectrolyte complex. Biomaterials 2008;29:4831e7.

103. Bae JW, Lee JH, Choi WS, et al. EPDIM peptide-immobilized porous chitosan beads for enhanced wound healing: preparation, characterizations and in vitro evaluation Mater Sci Eng 2009;29:697e701.

104. Sung JH, Hwang MR, Kim JO, et al. Gel characterisation and in vivo evaluation of minocycline-loaded wound dressing with enhanced wound healing using polyvinyl alcohol and chitosan. Int J Pharm 2010;392:232e40.

105. Yang C, Xu L, Zhou Y, et al. A green fabrication approach of gelatin/ CM-chitosan hybrid hydrogel for wound healing. Carbohydr Polym 2010;82:1297e305.

106. Li X, Chen S, Zhang B, et al. In situ injectable nanocomposite hydrogel composed of curcumin, N,Ocarboxymethyl chitosan and oxidized alginate for wound healing application. Int J Pharm 2012;437:110e9.

107. Li X, Nan K, Li L, et al. In vivo evaluation of curcumin nanoformulation loaded methoxy poly(ethylene glycol)- graft-chitosan composite film for wound healing application. Carbohydr Polym 2012;88:84e90.

108. Wijekoon A, Fountas-Davis N, Leipzig ND, et al. Fluorinated methacrylamide chitosan hydrogel systems as adaptable oxygen carriers for wound healing. Acta Biomater 2013;9:5653e64.

109. Balassa LL, Prudden JF. Application of chitin and chitosan in wound healing acceleration. In: Zikakis JP, editor. Chitin, chitosan and related enzymes. San Diego, CA: Academic Press; 1984. p. 296e305.

110. Muzzarelli RAA. Amphoteric derivatives of chitosan and their biological significance in chitin and chitosan. In: Skjak-Braek G, Anthonsen T, Sandford P, editors. London: Elsevier Applied Science; 1989, p. 87e99.

111. Minami S, Okamoto Y, Matsuhashi A. Application of chitin and chitosan in large animal practice. In: Brine CJ, Sandford PA, Zikakis JP, editors. Advances in chitin and chitosan. New York: Elsevier; 1992. p. 61e9.

112. Wang LS, Khor E, Wee A, et al. Chitosanealginate PEC membrane as a wound dressing: assessment of incisional wound healing. J Biomed Mater Res 2002;63:610e8.

113. Minami S, Okamoto Y, Hamada K, et al. Veterinary practice with chitin and chitosan. EXS 1999;87:265e77.

114. Muzzarelli RA, Mattioli-Belmonte M, Pugnaloni A. Biochemistry, histology and clinical uses of chitins and chitosans in wound healing. In: Jolles P, Muzzarelli RAA, editors. Chitin and chitinases. Basel: Birkhauser; 1999. p. 251e64.

115. Mi FL, Wu YB, Shyu SS, et al. Control of wound infections using a bilayer chitosan wound dressing with sustainable antibiotic delivery. J Biomed Mater Res 2002;59:438e49.

116. Burkatovskaya M, Castano AP, Demidova-Rice TN, et al. Effect of chitosan acetate bandage on wound healing in infected and noninfected wounds in mice. Wound Repair Regen 2008;16:425e31.

117. Kean T, Thanou M. Biodegradation, biodistribution and toxicity of chitosan. Adv Drug Deliv Rev 2010;62:3e11.

118. Halim AS, Lim CK. Biomedical-grade chitosan in wound management and its biocompatibility in vitro. In: Elnashar M, editor. Biopolymers. New York: Sciyo Publisher; 2010. p. 19e36.

119. Qi L, Xu Z. In vivo antitumor activity of chitosan nanoparticles. Bioorg Med Chem Lett 2006;16:4243e5.

120. Hu YL, Qi W, Han F. Toxicity evaluation of biodegradable chitosan nanoparticles using a zebrafish embryo model. Int J Nanomed 2011;6:3351e9.

121. Alnasser Y. Safety assessment of commonly used nanoparticles in biomedical application: impact on inflammatory processes. MSc thesis. MI,USA: Albany College of Pharmacy and Health Sciences, ProQuest LLC; 2013.

122. Miyazaki S, Yamaguchi H, Takada M. Pharmaceutical application of biomedical polymers: XXIX. Preliminary study on film dosage form prepared from chitosan for oral drug delivery. Acta Pharm Nord 1990; 2:401e6.

123. Chandy T, Sharma CP. Biodegradable chitosan matrix for the controlled release of steroids. Biomater Artif Cells Immobilization

Biotechnol 1991;19:745e60.

124. de Salamanca AE, Diebold Y, Calonge M, et al. Chitosan nanoparticles as a potential drug delivery system for the ocular surface: toxicity, uptake mechanism and in vivo tolerance. Invest Ophthalmol Vis Sci 2006;47:1416e25.

8

CHAPTER

N,N,N-TRIMETHYL CHITOSAN AND ITS POTENTIAL BACTERICIDAL ACTIVITY: CURRENT ASPECTS AND TECHNOLOGICAL APPLICATIONS

Débora P Facchi[1], Suelen P Facchi[1] and Alessandro F Martins[1,2]

[1]Environmental Engineering Program, Federal University of Technology, Paraná, Brazil

[2]Materials Science & Engineering Program, Federal University of Technology, Paraná, Brazil

ABSTRACT

N,N,N-Trimethyl chitosan (TMC) is one of most important chitosan derivative with stronger bactericidal property. Currently, the TMC has been attracted considerable attention, because it contains quaternized ammonium moieties (-+N(CH3)3) in its network. Its cationic property leads to a renowned bactericidal power. Besides, TMC has appropriate biocompatibility, biodegradability, mucoadhesive capacity and greater solubility at pH close to physiological condition regarding chitosan. In this way, there are not sources regarding revision papers toward TMC bactericidal action. So, this revision work reports the TMC potential to be applied in many fields, depicting its bactericidal performance, as well as, its bactericidal mechanism and some current technological aspects that must be considered.

INTRODUCTION

Chitosan (a chitin derivative) is water-insoluble (at physiological pH condition) and in most organic solvents, except in organic acids, which substantially limits its usefulness [1]. On the other hand, the N,N,Ntrimethyl chitosan is the most famous chitosan derivative, obtained from reductive methylation of amino moieties. Several methodologies concerning the TMC synthesis are depicted in the (Figure 1). Among these, the highlighted methodologies were proposed by Sieval et al., Verheul et al., De Brito and Assis, and Benediktsdottir et al. (Figure 1) [2-5].

The most commonly used synthesis is based on the amino groups reduction (from two steps) in strong base, sodium iodide and iodomethane presence, using N-methyl-2-pyrrolidinone (NMP) as solvent [2]. However, this pathway provides the -OH/-NH2 uncontrolled methylation, being its reproducibility a negative factor (Figure 1). Among these procedures, the most important and simple synthesis methodology has been described [3]. This route offers lower cost and exhibits excellent reproducibility and such pathway occurs from two steps. In the first step, the N,N-dimethyl chitosan (DMC) is performed from formic acid-formaldehyde mixture (Eschweiler- Clarke) at 70°C (Figure 1). Then, the fully N-dimethylated DMC is quaternizaed, using only iodomethane under NMP presence and sodium iodide and sodium hydroxide absences (Figure 1). This methodology leads to TMC free of O-methylation. De Brito and Assis [4] developed another methodology using the dimethylsulfate as reductor agent instead iodomethane (Figure 1). On the other hand, Bendiktsdottir et al. [5] synthesized a fully quaternized TMC from protection strategies toward –OH groups (using the tertbutyldimethylsilyl - TBDMS protection group and tetrabutylammonium fluoride (TBAF) as desprotetor agent), avoiding O-methylation (Figure 1). The Figure 1 will be further discussed in the following sections. TMC has received great attention in the last 15 years, due to its excellent properties of biodegradability, biocompatibility, mucoadhesion, excellent water-solubility, and especially for its antimicrobial potential [6]. TMC may exhibit bactericidal activity up to 700 times greater than the neat chitosan. This property is intrinsically linked to -+N(CH3)3 quaternary groups existence in its network and some studies show that larger N-quaternized fraction, better bactericidal performance is reached i.e., the higher TMC quaternization degree (DQ) promotes higher bactericidal power [7]. Factors as molar weight [8], specie types toward -+N(CH3)3 sites (counterion types) [9], synthesis methodology, as well as, the adopted purification process [10], and culture assay conditions (temperature and ionic strength) [8] can change the TMC antimicrobial capacity.

Some of these factors have been extensively studied and they are not reported in this opportunity. The effects promoted by these factors regarding bactericidal activity were recently revised in a work published by Hosseinnejad and Jafari [8], they will not be discussed again.

Recent studies reinforce that TMC can be applied in pharmaceutical and medical field [11-13], as well as, in food industry [1]. TMC basedmaterials (thin films, beads, thermosensitive hydrogels, polyeletrolyte complexes, nanoparticles, etc.,) may be designed as drug delivery matrices [13], as building devices coated with bactericidal and cell growth potentials [12,14] and developing biodegradable membranes [15,16] with bactericidal action for use in food industry, among others. This paper highlights the latest issues concerning TMC antimicrobial potential, as well as, some TMC-based technological applications.

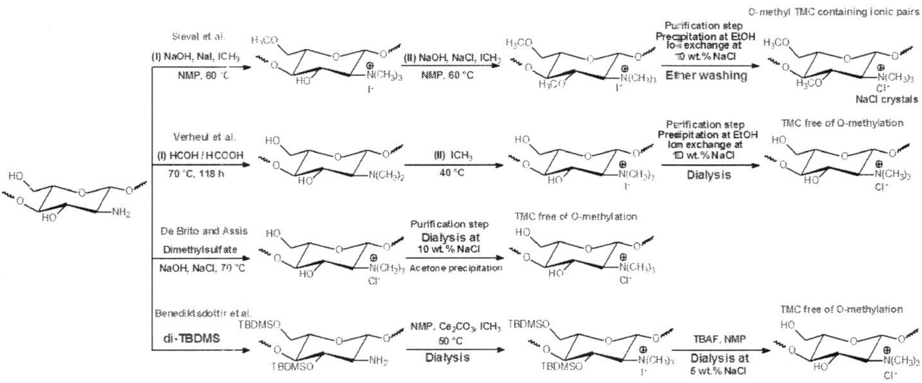

Figure 1. The most important routes for the TMC synthesis [2-5].

Probably TMC Bactericidal Mechanism

TMC bactericidal activity is the most targeted parameter toward the pharmacy and medicine fields. TMC showed potential to kill several bacteria types, such as Escherichia coli [9,10], Staphylococcus aureus [9], Pseudomonas aeruginosa [17], Listeria innocua [18] and Enterococcus facialis [17], i.e., it has antimicrobial activity on gramnegative and gram-positive bacteria. On the other hand, the precise TMC bactericidal mechanism is difficult to identify, because such activity depends on many parameters, as previously reported section [6,8-10]. Studies indicate that TMC has potential against Gram negative and Gram-positive bacteria, and its action can be better understood from analysis as shown in Figure 2. Gram-negative cells are composed by a multilayer cell surface containing an outer membrane, a peptidoglycan layer between periplasmic space and an internal cytoplasmic membrane (Figure 1) [6,9]. On the other hand, gram-positive membrane cells are constituted by a broad dense wall that consists of 15–40 interconnecting peptidoglycans layers and an internal membrane cell [6,9]. Anionic phospholipid dipalmitoylphosphatidylglycerol (PDPPG) is the major component of both membrane cells (gram-negative and gram-positive bacteria) (Figure 1) [6].

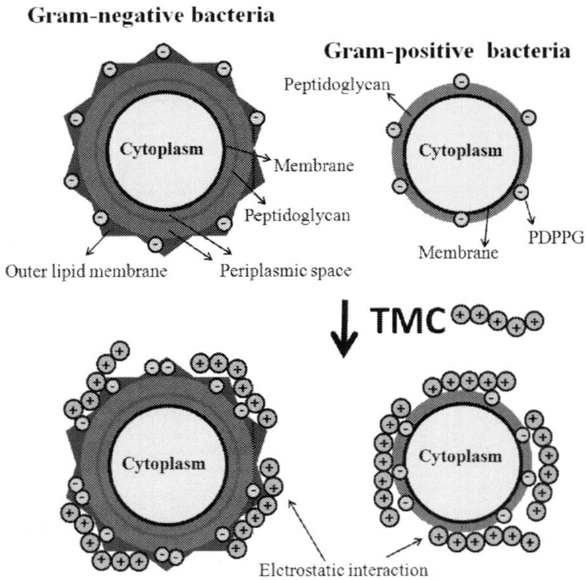

Figure 2. Bactericidal mechanism of TMC [6,9].

Besides, outer membrane (OM) on gramnegative bacteria contains polyanionic lipopolysaccharides (PLPS), which are stabilized by

Mg2+ and Ca2+ divalent ions [6]. Such species bind on out membrane, maintaining them stables, being essential to OM integrity. Therefore, gram-negative membrane cells have more negative charge densities than gram-positive cells.

Cationic TMC polymer has higher positive charge densities that can interact with the cell walls (Figure 1) [9,10]. This interaction occurs among cationic N-quaternized moieties (-+N(CH3)3) on TMC networks and anionic molecules, such as PDPPG on bacteria cells (Figure 1). It leads to a large increase on membrane permeability, causing structural changes (distortion-disruption), as well as, a severe leakage of cytoplasm constituents and eventually results in the death of bacteria [19]. TMC has greater adsorption capacity onto bacteria cell surface, especially on gram-negative bacteria cell, due its higher negative charge property, regarding the gram-positive bacteria [6].

Current Aspects Concerning the Bactericidal Potential of TMC

Some important points regarding the TMC bactericidal potential cannot be unnoticed. A recently paper showed that adopted purification process on TMC synthesis significantly influence its bactericidal activity [10]. TMCs are obtained as iodide salts and then they are dissolved in a 10% wt. sodium chloride solution for ion exchange (Figure 1). TMC chloride is more water-soluble than TMC iodide due to smaller chloride ionic radius concerning the iodide [2].

Then, the TMC chloride aqueous solution is precipitated in ether and then it is centrifuged and washed/purified. However, TMC chloride can be purified by two ways (Figure 1). i) After centrifugation, the TMC chloride is washed several times with ether solvent and dried at 40°C to finally be obtained as a white crystalline powder [10]; ii) After centrifugation the TMC can be dispersed again in water and dialyzed (for 3 to 4 days), frozen and lyophilized [10]; The second process is more efficient and it has low cost, because it is not necessary carry out the TMC washing, using organic solvent. In the TMC chloride precipitation process (in EtOH), it attracts to its structure sodium chloride crystals, an insoluble salt in ethanol and ether solvents. Upon precipitation, sodium chloride crystals are directed to the -+N(CH3)3Cl- group, leading the 'ionic pairs' formation (Figure 1) [10]. The ionic pairs between TMC chloride/NaCl crystals drastically decrease the TMC bactericidal potential, whereas they are shielding the TMC N-quaternized groups, reducing the potential interaction of these sites with the bacteria cell membranes [10]. So, the dialysis may be used to remove the NaCl from TMC structure, purifying it (Figure 1). After dialysis, TMC chloride must be frozen and lyophilized to be obtained in an amorphous form [10]; feature of polysaccharide macromolecules.

TMC chlorides obtained by our research group was carried out according to the methodologies proposed by Sieval et al. [2] and Verheul et al. [3], using some modifications on experimental procedures [10]. TMC chloride O-methylated and TMC chloride free of O-methylation, containing the same DQ (≈15%) were evaluated, regarding their bactericidal potentials against Escherichia coli (E. coli) (ATCC 26922) [10]. In this case, TMC chlorides were produced atdialysis absence, i.e., TMCs with high crystallinity were performed due to ionic pairs' formation [10], as depicted in the reaction at the top on Figure 1. TMC salts were also successful synthesized and purified from dialysis method, i.e., amorphous TMCs free of ionic pairs were synthesized [10].

TMCs free of ionic pairs (purified by dialysis) showed stronger antimicrobial action against E. coli in only 6 h. Both O-methyl TMC and TMC free of O-methylation (both 15% DQ) have more than 90% inhibition. On the other hand, regarding the TMCs containing ionic pairs, their bactericidal performance does not exceed 40% inhibition [10]. These results showed that TMC should be purified by dialysis process to enable its bactericidal performance. Another recent study depicted that TMC counterion species (counterions on -+N(CH3)3 sites) dramatically influences the antimicrobial effect [9]. Several TMC salts (TMC iodide, TMC bromide, TMC chloride, TMC acetate and TMC sulfate) with 60% quaternization degree were obtained and their microbial potentials investigated against E. coli (ATCC 25922) and Staphylococcus aureus (S. aureus) (ATCC 25923) [9]. The salts containing delocalized π counterions have better bactericidal activities toward both bacteria (E. coli and S. aureus). TMC iodide and TMC bromide showed weak potential to kill bacteria, due to the higher ionic volume (cm3 mol-1) and ionic radius (pm), as well as, the delocalized π absence for the Brand I- counterions [9]. In order, the bactericidal effect followed the sequence: TMC sulfate>TMC acetate>TMC chloride>TMC bromide ≈TMC iodide [9]. So, TMC sulfate and TMC acetate presented the highest activities against both bacteria, whereas the number of - +N(CH3)3 groups "available" to interact with the bacteria cell membranes should be larger [9].

Most Recent TMC Applications

TMC-based materials have been much attention in these last years, mainly for the pharmaceutical and medical fields. The cationic property of TMC allows its association with polyanionic macromolecules, like heparin, sodium alginate, among others [20,21]. TMC/heparin multilayer thin films carried out from the layer-by-layer assembly [20,22]. Such assembled film prepared with O-methyl TMC, containing 80% DQ showed stronger bactericidal action (64.6% inhibition at 6 h) against E. coli (ATCC 26922) at physiologic pH condition (7.4) [22]. Sajomsang et al. [23] revealed that N,N,Ntrimethyl ammonium group showed high bactericidal activity toward E. coli (ATCC 25922) and S. aureus (ATCC 6538) [23].

Hydrogel beads based on TMC/alginate polyelectrolyte complexes showed bactericidal property against E. coli (ATCC 26922) cells, mainly when associated with silver nanoparticles [13]. In this case, these metallic nanomaterials enhanced TMC/alginate bactericidal performance to 91% inhibition at 24 h (pH 7.4) [13]. Quaternary ammonium TMC prepared from amino group moieties, exhibited mproved water solubility and stronger bactericidal activity over an entire pH range. So, quaternized sites can increase the potential biomedical applications on anti-infection [19]. Electrospun membranes based on TMC/poly(vinyl pyrrolidone) had high antibacterial activity against S. aureus (749) and E. coli (3588) [15].

Another electrospun nanofibers, formulated from association of poly(e-caprolactone) coated by poly(acrylic acid)/TMC iodide polyelectrolyte complex, inhibited the growth of S. aureus and E. coli, as well as, suppressed the pathogenic bacteria adhesion of S. aureus [16]. Nanoparticles containing quaternized moieties on their surfaces also suppressed the E. coli growth [24]. Many TMC-formulations have been conducted and designed for technological applications, due to bactericidal action of these based-materials.

Future Trends

TMC is famous because it not only retains the chitosan properties, such as biocompatibility, biodegradability and non-toxicity, but also it can enhance the antimicrobial activity, water-solubility and mucoadhesion capacity [22]. Therefore, the TMC has attracted much attention and recent studies show its wide application range.

Zhou et al. [12] obtained TMC fibers (varying the TMC quaternization degree) with high water absorption performance and excellent bactericidal activity (than chitosan fibers) against E. coli (>63% inhibition) and S. aureus (>99% inhibition). TMC fibers regarding chitosan fibers could significantly enhance the contraction and wound re-epithelialization toward the mouse embryo fibroblast cells, leading to the wound dressing potential [12]. TMC-materials showed higher water absorbance capacity and stronger bactericidal effect, suppressing excessive would maceration i.e., the eschar formation and chronic clinical events at only 12 days [12, 14]. So, TMC fiber has potential to be used as components for wound healing performance. Currently, TMC and TMC-based materials are being applied to obtain bactericidal materials. These are designed for orthopedic application [19], formulation of dental resin [11], repair engineering and wound healing/cell growth [12,14] and food industry [1,8], among others.

TMC has attracted attention on food industry due to its easy acquisition, preparation low cost and mainly due to its extraordinary bactericidal activity. Currently, there is a great race for the new technologies development, in order to obtain bactericidal films, aiming new food preservative strategies. In this way, TMC based film can be used to kill E. coli (O157:H7), one of the most infamous foodborne pathogen that has been detected in various foods, including cheese, neat milk, undercooked meat, and spinach [1].

Chitosan is the most bactericidal agent applied like food additive/ preservative [1]. However, the chitosan has lower bactericidal activity than TMC and chitosan based-materials in solid-state did not show considerable bactericidal action [6]. In this sense, the TMC synthesis may become a reality on industrial scale. The food industry has continually sought the development of bactericidal films that may increase the lifetime of food packaging. In this case, the TMC must be used for this purpose, in order to obtain bactericidal films with excellent performance for food preservation.

REFERENCES

1. Kong M, Chen XG, Xing K, Park HJ (2010) Antimicrobial properties of chitosan and mode of action: a state of the art review. Int J Food Microbiol 144: 51-63.
2. Sieval AB, Thanou M, Kotze′ AF, Verhoef JC, Brussee J, et al. (1998) Preparation and NMR characterization of highly substituted N-trimethyl chitosan chloride. Carbohydr Polym 36: 157-165.
3. Verheul RJ, Amidi M, van der Wal S, van Riet E, Jiskoot W, et al. (2008) Synthesis, characterization and in vitro biological properties of O-methyl free N,N,N-trimethylated chitosan. Biomaterials 29: 3642-3649.
4. De Britto D, Assis OBG (2007) A novel method for obtaining a quaternary salt of chitosan. Carbohydr Polym 69: 305-310.
5. Benediktsdóttir BE, Gaware VS, Rúnarsson ÖV, Jónsdóttir S, Jensen KJ, et al. (2011) Synthesis of N,N,N-trimethyl chitosan homopolymer and highly substituted N-alkyl-N,N-dimethyl chitosan derivatives with the aid of di-tert-butyldimethylsilyl chitosan. Carbohydr Polym 86: 1451-1460.
6. Martins AF, Facchi SP, Follmann HD, Pereira AG, Rubira AF, et al. (2014) Antimicrobial Activity of Chitosan Derivatives Containing NQuaternized Moieties in Its Backbone: A Review. Int J Mol Sci 15: 20800-20832.
7. Xu T, Xin M, Li M, Huang H, Zhou S, et al. (2010) Synthesis, characteristic and antibacterial activity of N,N,N-trimethyl chitosan and its carboxymethyl derivatives. Carbohydr Polym 81: 931-936.
8. Hosseinnejad M, Jafari SM (2016) Evaluation of different factors affecting antimicrobial properties of chitosan. Int J Biol Macromol 85: 467-475.
9. Follmann HD, Martins AF, Nobre TM, Bresolin JD, Cellet TS, et al. (2016) Extent of shielding by counterions determines the bactericidal activity of N,N,N-trimethyl chitosan salts. Carbohydr Polym 137: 418-425.

10. Martins AF, Facchi SP, Follmann HD, Gerola AP, Rubira AF, et al. (2015) Shielding effect of 'surface ion pairs' on physicochemical and bactericidal properties of N,N,N-trimethyl chitosan salts. Carbohydr Res 402: 252-260.
11. Song R, Zhong Z, Lin L (2016) Evaluation of chitosan quaternary ammonium salt-modified resin denture base material. Int J Biol Macromol 85: 102-110.
12. Zhou Z, Yan D, Cheng X, Kong M, Liu Y, et al. (2016) Biomaterials based on N,N,N-trimethyl chitosan fibers in wound dressing applications. Int J Biol Macromol 89: 471-476.
13. Martins AF, Monteiro JP, Bonafé EG, Gerola AP, Silva CTP, et al. (2015) Bactericidal activity of hydrogel beads based on N,N,N-trimethyl chitosan/alginate complexes loaded with silver nanoparticles. Chin Chem Lett 26: 1129-1132.
14. Fan L, Yang J, Wu H, Hu Z, Yi J, et al. (2015) Preparation and characterization of quaternary ammonium chitosan hydrogel with significant antibacterial activity. Int J Biol Macromol 79: 830-836.
15. Ignatova M, Manolova N, Rashkov I (2007) Novel antibacterial fibers of quaternized chitosan and poly(vinyl pyrrolidone) prepared by electrospinning. Eur Polym J 43: 1112-1122.
16. Kalinov K, Ignatova M, Maximova V, Rashkov I, Manolova N (2014) Modification of electrospun poly(ε-caprolactone) mats by formation of a polyelectrolyte complex between poly(acrylic acid) and quaternized chitosan for tuning of their antibacterial properties. Eur Polym J 50: 18-29.
17. Rúnarsson ÖV, Holappa J, Malainer C, Steinsson H, Hjálmarsdóttir M, et al. (2010) Antibacterial activity of N-quaternary chitosan derivatives: Synthesis, characterization and structure activity relationship (SAR) investigations. Eur Polym J 46: 1251-1267.
18. Belalia R, Grelier S, Benaissa M, Coma V (2008) New Bioactive Biomaterials Based on Quaternized Chitosan. J Agric Food Chem 56: 1582-1588.
19. Tan H, Ma R, Lin C, Liu Z, Tang T (2013) Quaternized chitosan as an antimicrobial agent: antimicrobial activity, mechanism of action and biomedical applications in orthopedics. Int J Mol Sci 14: 1854-1869.

20. Follmann HD, Naves AF, Martins AF, Félix O, Decher G, et al. (2016) Advanced fibroblast proliferation inhibition for biocompatible coating by electrostatic layer-by-layer assemblies of heparin and chitosan derivatives. J. Colloid Interface Sci 474: 9-17.
21. Martins AF, Follmann HD, Monteiro JP, Bonafé EG, Nocchi S, et al. (2015) Polyelectrolyte complex containing silver nanoparticles with antitumor property on Caco-2 colon cancer cells. International Journal of Biological Macromolecules 79: 748-755.
22. Follmann HD, Martins AF, Gerola AP, Burgo TA, Nakamura CV, et al. (2012) Antiadhesive and Antibacterial Multilayer Films via Layer-by- Layer Assembly of TMC/Heparin Complexes. Biomacromolecules 13: 3711-3722.
23. Sajomsang W, Ruktanonchai UR, Gonil P, Warin C (2010) Quaternization of N-(3-pyridylmethyl) chitosan derivatives: Effects of the degree of quaternization, molecular weight and ratio of N-methylpyridinium and N,N,N-trimethyl ammonium moieties on bactericidal activity. Carbohydr Polym 82: 1143-1152.
24. Wiarachai O, Thongchul N, Kiatkamjornwong S, Hoven VP (2012) Surface-quaternized chitosan particles as an alternative and effective organic antibacterial material. Colloids Surf, B 92: 121-129.

9

CHAPTER

GANCICLOVIR LOADED CHITOSAN NANOPARTICLES: PREPARATION AND CHARACTERIZATION

Patel R[1], Gajra B[1,2], Parikh RH[1] and Gayatri Patel[1]

[1]Ramanbhai Patel College of Pharmacy, Charotar University of Science and Technology, CHARUSAT Campus, Changa – 388421, Anand, Gujarat, India

[2]Walmart Pharmacy, 3600 Majormeckanzy drive, Vaughan, Ontario, Canada

ABSTRACT

Chitosan Nanoparticles (CSNPs), as drug carrier, can be utilized for enhancing permeability of poorly absorbed drugs. The aim of this work was to enhance the permeability of Ganciclovir (GCV) by loading into CSNPs. An ionic gelation method was undertaken to develop GCV loaded CSNPs. Several process and formulation parameters were screened and optimized through 25-2 fractional factorial design and Box-Behnken design respectively. The CSNPs were evaluated and characterized for their particle size and shape, surface charge, entrapment efficiency, crosslinking mechanism (dried CSNPs) and drug release study. The optimized CSNPs were found with particle size of 121.20 ± 2.7 and entrapment efficiency (%EE) of 85.15 ± 1.1%. Transmission electron microscopy, scanning electron microscopy and dynamic light scattering technique revealed spherical particles with uniform size. The *in vitro* release profile was found to be sustained up to 24 hr. Thus, incorporation of GCV into CSNPs results in enhanced permeability, that may in turn increase overall oral absorption of the drug.

INTRODUCTION

Ganciclovir (GCV) (9-(1,3-dihydroxy-2-propoxymethyl) guanine) (DHPG), is a BCS class III drug. GCV (log P=-1.7) having molecular formula C9H13N5O4 and molecular weight 255.2 g/mol. It is the first antiviral drug that proved to be efficacious in the treatment of Cytomegalovirus (CMV) disease in humans [1,2]. GCV is also used for maintenance therapy and prophylaxis of CMV. Oral bioavailability of GCV is ~5%. Possible reasons for the poor bioavailability are poor permeability through the gastro-intestinal tract (GIT) and P-gp substrate activity [3].

Chitosan (CS), an N-deacetylation product of chitin, a natural polysaccharide, is mostly found in the exoskeleton of crustaceans, insects, and fungi [4]. After deacetylation process, CS is able to dissolve in acidic medium and becomes the only polysaccharide that possesses high density of positive charges, due to the protonation of amino groups on its backbone. Besides this unique characteristic, CS has been proved to have many other intrinsic properties, such as non-toxicity, biocompatibility and biodegradability [5]. It is mucoadhesive polymer and can increase the residence time at the site of absorption and has favourable controlled drug-release abilities [6]. Chitosan has a low level of toxicity where oral LD50 value is found to be in excess of 16 g/kg body weight of mouse [4]. It has been demonstrated that CS (when protonated) affects cell permeability and enhances paracellular permeability of drugs across the mucosal epithelia by opening the intercellular tight junctions.

Chitosan causes reversible effects on epithelial morphology and temporarily alters gating properties of tight junctions. Juan M Irache, 2001 has done work on GCV -loaded albumin

Nanoparticles. They found the albumin carriers were able to release GCV in a sustained way after incubation between the drug and the protein, prior the preparation of Nanoparticles [7]. Solid lipid Nanoparticles were prepared by Gang Cheng, 2013 [8]. They found that SLN modified with borneol is a potential delivery system for transporting GCV to the central nervous system. In-vitro antiviral efficacy of the GCV complexed with-cyclodextrin on human cytomegalovirus clinical strains was performed by Chantal Finance, 2002 [9]. They found that the complexed GCV was more effective than free GCV against all Human cytomegalovirus strains tested. Gao Li, 2011 examined the effects of some common excipients on the intestinal absorption of GCV [3]. Enhancements of intestinal absorption of GCV by Pluronic F-68, verapamil, Polyethylene glycol-400, Tween-80 and Cremophor EL-35 excipients are probably due to inhibition of P-gpmediated drug efflux. Long-circulating liposome-encapsulated GCV is a new approach to drug carriers to enhance the efficacy of suicide gene therapy was concluded by Yoshie Maitani, 2007 [10]. Polymeric Nanoparticles (NPs) used for drug delivery is defined as colloidal systems made of solid polymers. They offer a significant improvement over traditional oral and intravenous methods of administration in terms of efficiency and effectiveness [11]. They may be classified according to their size and the processes of preparation. Nanocapsules are composed of a polymeric wall containing inner core filled with a liquid where the drug is entrapped and nanospheres are made of a solid polymeric matrix in which the drug can be dispersed. Drug substances may be either adsorbed at the surface of the polymer or encapsulated within the particle. Particles may be produced by polymerization of synthetic monomers, or dispersion of synthetic polymers or natural macromolecules [12]. Polymeric NPs are made from biocompatible and biodegradable materials like polymers either natural like cellulose, CS, pullulan, gelatin, alginate or synthetic like poly-lactide (PLA), polylactide co-glycolide (PLGA), poly-ε-caprolactone (PCL). NPs can be taken up by the intestinal epithelial cells via different mechanisms like transcellular transport which involve the vesicular transport through M cells of Peyer's patches, transport via the epithelial cell lining (enterocytes) in the intestinal mucosa and by paracellular pathway [13]. The present study was undertaken to develop GCV loaded CSNPs with the aim of enhanced permeability. GCV loaded CSNPs were prepared by ionic gelation method and evaluated for physicochemical properties. CS-TPP complex was formed by the ionic interaction between positively charged CS and negatively charged phosphoric ions of TPP.

The mechanism of cross-linking of CS with TPP could be either by deprotonation or ionic interaction [14]. In-vitro drug release studies were performed in order to elucidate the ability of the developed formulation to release GCV. *Ex-vivo* permeability study was performed by normal sac method.

MATERIALS AND METHODS

Materials

Ganciclovir (GCV) was kindly gifted by Bakul Finechem Research Center, Mumbai. Chitosan (molecular weight=110 kDa, 80.0% deacetylation degree) was gratis sample from Chitopharm S, Norway (USA). Sodium Tripolyphosphate (Cross-linking agent) was purchased from Sigma Aldrich, Mumbai (India). The water used was pre-treated with the Milli-Q plus system (Millipore, Q-5 UVS, India). All other materials were of analytical grade were used.

Preparation of GCV loaded CSNPs

The GCV loaded CSNPs were prepared by ionic gelation method. The possible mechanism for ionic gelation (Figure 1) is that the ionic interaction between positively charged CS with negatively charged polyanion Sodium Tripolyphosphate (TPP) [15]. Briefly in the preparation of optimized batch, CS solutions were prepared by dissolving CS into Milli-Q® water containing glacial acetic acid (2%).

GCV was dissolved in Milli-Q® water containing TPP. NPs were formed by adding TPP solution drop wise onto a CS solution under mechanical stirring (Remi, India) at 1000 rpm for 100 min. Thereafter, the CS-TPP particle suspension was processed under ultrasonication by Probe sonicator (VCX- 500, Vibra cell, U.S.A.) for 5 minutes, producing CSNPs with controlled particle sizes. The dispersion was stored in refrigerator until further evaluation. The prepared GCV loaded CSNPs were dried by fluidized bed drying method [16].

Experimental Design

From the prior experience and literature survey, the process and formulation parameters were identified and screened through 25-2 fractional factorial design. Eight batches were prepared and evaluated for particle size, percentage entrapment efficiency (%EE) and morphology. Table 1 shows the independent and dependent variables with their coded and actual values for 25-2 Fractional Factorial Design. Optimize the identified factors from the screening design, Box- Behnken Design was used with 15-run, 3-factor, 3-level was used. This design is suitable for investigating the quadratic response surface and for constructing a second order polynomial model. The dependent and independent variables with their coded and actual values for Box- Behnken Design are shown in the Table 2. Based on the results of the screening design sonication time (5 min), stirring speed (1000 RPM) and drug amount (75 mg) was considered as constant variables for the preparation of the formulation.

Morphology Observation

Morphology of the formulation was determined by Transmission electron microscopy (TEM) (Technai20, Philips, Holland) by placing drops of CSNPs dispersion on Lanthanum Hexabromide (LaB6) grid and allow to air dried. The NPs were viewed under a high resolution microscope at 200 kV accelerating voltage [17]. Scanning electron microscopy (SEM) (JSM 6010 LA, JEOL, USA) was used to access the morphology/shape of the NPs. A sample of dried CSNPs was placed on a double stick tape over aluminium stubs to get a uniform layer of particles. Sample was platinum coated for 20 sec. Then the sample was observed by SEM at 10 kV [18].

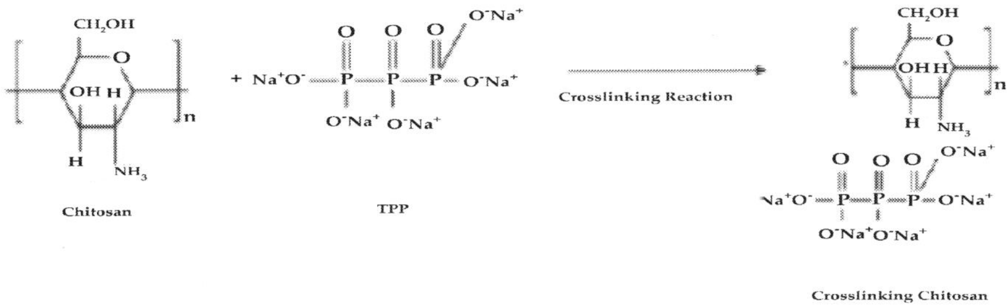

Figure 1 Mechanism of crosslinking between Chitosan and TPP

Table 1 Variables with their coded and actual values for 25-2 fractional factorial design.

Independent variables	Low	High
Coded values	(-1)	(1)
A=Drug: polymer ratio (w/w)	1:1.5	1:3
B=Concentration of TPP (%w/v)	0.2	0.4
C=Stirring time (min)	60	120
D= Sonication time (min)	5	10
E=Stirring Speed (rpm)	500	1000
Dependent variables Constrains		
Y_1=Particle size (nm) 100-200 nm Y_2=Entrapment Efficiency (%) Maximum Y_3 Morphology Spherical with uniform size		

Table 2 Variables with their coded and actual values for Box-Behnken design.

Independent variables	Low	Medium	High
Coded values	(-1)	(0)	(1)
A=Drug: polymer ratio (w/w)	1:2	1:3	1:4
B=Concentration of TPP (%w/v)	0.3	0.4	0.5
C=Stirring time (min)	50	100	150
Dependent Variables Constrains			
Y_1=Particle size (nm) 100-200 nm Y_2=Entrapment Efficiency (%) Maximum			

Particle Size and Zeta Potential

The size of the particles was determined by dynamic laser scattering technique using Malvern nano S90 (Malvern Instruments, UK) particle size analyzer at 25°C with an angle of 900. The zeta potential is crucial parameter for stability in aqueous nanoparticulate dispersion. The zeta potential measures the surface charge of the particles. The pellet obtained after centrifugation of the nanoparticulate dispersion was redispersed with water. The diluted sample was taken in a capillary type cell and the zeta potential was determined by Zetasizer Nano ZS (Malvern Instruments, UK) [19].

Entrapment Efficiency (%EE)

% EE of the drug was determined by High Performance Liquid Chromatography (HPLC) (LC-2010C HT, Shimadzu, Japan). Formulations were centrifuged at 10,000 rpm for 30 min. Supernant was collected and analyzed for drug content at 254 nm by HPLC [20]. The %EE was calculated as follows:

%EE=(Sa-Sb)/Sa*100

Where, Sa is the total amount of drug in system, Sb is the amount of drug in supernatant after centrifugation.

Fourier Transforms Infrared (FTIR) Spectroscopy

FTIR spectrum was recorded using NICOLET-6700 (Thermo Scientific, US) to confirm the cross-linking reaction between the phosphoric group of the TPP and amino group of CS. The pellets were prepared by homogenously dried formulation in dried KBr in a mortar and pestle and then powder was compressed under vacuum using round flat face punch to produce pellet compact [20]. The sample was placed in IR light path and spectra were scanned over the wavelength number range of 4000 to 400 cm-1 [21].

Powder X-ray Diffraction (PXRD) Study

PXRD analysis provides the crystal lattice arrangements and gives the information regarding the degree of crystallinity in the formulation. It is also used for the identification of physical state of drug in the formulation. Powder X-ray diffraction spectra of dried CSNPs were recorded at room temperature using X-ray diffractometer (D2 PHASER, Bruker, Germany) with a voltage of 3 kV, 5 mA current, 40/min scanning speed. The samples were scanned from 0 to 600 (2θ) range with a step size 0.030 and a step interval of 0.1 second (sec) [22].

Conductivity Study

The conductivity study was performed to confirm the crosslinking reaction by conductivity meter (Systronics; 307). The change in conductivity was measured after each mL of addition of TPP solution to the CS solution [23].

In-vitro Drug Release Study

In-vitro drug release study of CSNPs dispersion was carried out in diffusion cell apparatus (J-FDC-07, Orchid Scientifics and Innovations India Pvt. Ltd.) in phosphate buffer pH 6.8. At predetermined time intervals the samples was withdrawn and replenish with fresh medium and the absorbance was measured by HPLC (LC-2010C HT, Shimadzu, Japan) at 254 nm. Data obtained from the in-vitro drug release for formulation in different release medium were fitted to various kinetic models. Each experiment was performed in triplicate. The drug release mechanism and linearization were determined by finding the goodness of fit (R2) and sum squared of residuals (SSR) for each kinetic model [24].

RESULTS AND DISCUSSION

From the literature, various process and formulation parameters were identified and screened through 25-2 fractional factorial design. The selected parameters were then optimized through Box-Behnken design.

Statistical Analysis of Experimental Data

The results of the experimental design were analyzed which provided considerable useful information and reaffirmed the utility of statistical design for the conduct of experiments. Table 3 shows the 25-2 Fractional Factorial Design with results. From the selected independent variables, including the drug to polymer ratio, concentration of TPP and the stirring time, significantly influenced the observed responses; particle size and %EE (%).

Table 3 The 25-2 Fractional Factorial Design with results (mean n =3 + SD)

Batch Code	Results		
	Y_1	Y_2	Y_3
G1	364.9 ± 4.7	67.83 ± 1.4	Spherical
G2	298.3 ± 6.4	63.94 ± 2.1	Spherical
G3	257.4 ± 3.6	73.20 ± 1.2	Spherical
G4	348.6± 11.2	60.04 ± 4.8	Spherical
G5	276.3 ± 3.3	69.39 ± 5.1	Spherical
G6	310.8± 15.3	59.10 ±5.4	Spherical
G7	186.9 ± 2.5	80.81 ± 2.1	Spherical

Figure 2 shows the Pareto chart of main effects of 25-2 fractional factorial design. The height of each bar gives the information about the significance of the variables. Here, Drug: polymer ratio, concentration of TPP and stirring time was observed with significant effects. This can also be confirmed by the 3D surface plot (Figure 3) for the main effects obtained from 25-2 fractional factorial design. For the optimization of formulation Box-Behnken design was applied. All the batches were formulated and evaluated for the particle size and %EE. The results of all the batches are shown in Table 4.

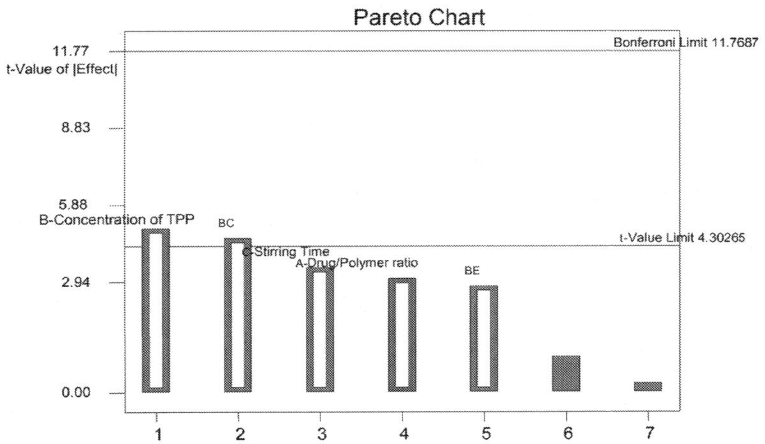

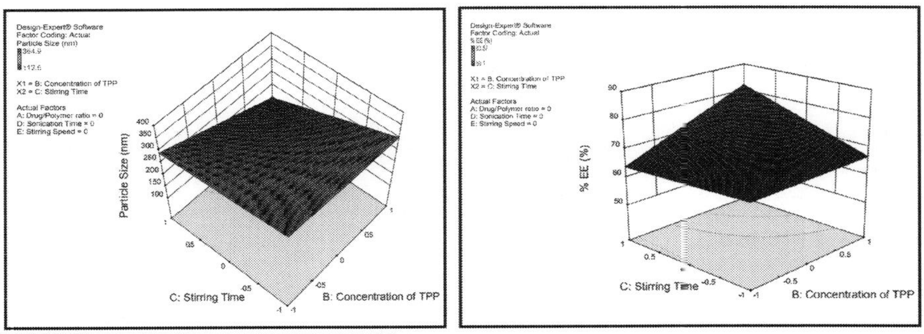

Figure 2 Pareto Chart of main effects obtained from 25-2 fractional factorial design.

Particle Size

The particle size for all the batches was found between 121.20 ± 2.7 nm to 294.5 ± 15.4 nm. The polynomial equation for particle size is as follows: Particle size [Y1]=+ 121.53 -17.54A (P<0.05) – 0.13B (P<0.05) – 4.39C (P>0.05) -11.79 AB (P<0.05) - 9AC (P<0.05) + 19.93 BC (P<0.05) + 86.16A2 (P>0.05) + 40.18B2 (P>0.05) – 66.06C2 (P> 0.05) (2)

Among all the independent variables selected, A, B, C and AB, AC and BC were significant model terms as the p<0.05. Here, variable A, B and C have negative effect on particle size as revealed by negative value of coefficient in the Equation (2). There is increase in the particle size due to increase in the concentration of TPP would be due to crosslinkage between TPP and CS. This result is also confirmed by the response surface 3D plot for the particle size in Figure 4.

Batch Code	Dependent variables	
	Y_1	Y_2
G1	212.3 ± 3.1	68.87 ± 1.7
G2	273.6 ± 10.9	59.95 ± 0.5
G3	245.7 ± 4.6	64.75 ± 2.1
G4	259.9 ± 6.9	70.14 ± 2.8
G5	235.0 ± 3.7	72.21 ± 3.7
G6	285.4 ± 25.3	78.54 ± 0.7
G7	280.1 ± 30.15	71.91 ± 3.9
G8	294.5 ± 15.4	65.33 ± 2.4
G9	275.2 ± 10.8	69.45 ± 1.6
G10	225.0 ± 12.4	82.17 ± 2.5
G11	190.7 ± 22.7	79.62 ± 2.9
G12	220.2 ± 3.8	81.24 ± 1.5
G13	121.3 ± 1.2	84.00 ± 2.1
G14	120.7 ± 1.9	85.15 ± 1.1
G15	122.6 ± 2.1	82.91 ± 1.8

%EE

The %EE for all the batches was found to between 59.95 ± 0.5 and 85.15 ± 1.1%. The polynomial equation for %EE is as follows: %EE=+84.02 – 0.47A (P<0.05) + 2.55B (P<0.05) – 0.53 C (P<0.05) + 3.58AB (P<0.05) – 3.23AC (P<0.05) - 2.78BC (P>0.05) - 12.11A2 (P>0.05) - 5.98B2 (P>0.05) + 0.085C2 (P>0.05) (3). From the Equation (3), it can be found that all the independent variables selected A, B, C and AB, AC were significant model terms as the p-value is less than 0.05. Here, variable A and C have negative effect on particle size as revealed by negative value of coefficient in the Equation (3). There is decrease in the %EE with increase in the drug: polymer ratio. This may be due to the fact that at higher concentration CS molecules in the dispersion are present very close to each other during precipitation by polyanion. So, they got precipitated to form a large particle. And this larger particle is having low entrapment of drug.25 This result is also confirmed by the response surface 3D plot for the particle size in Figure 5.

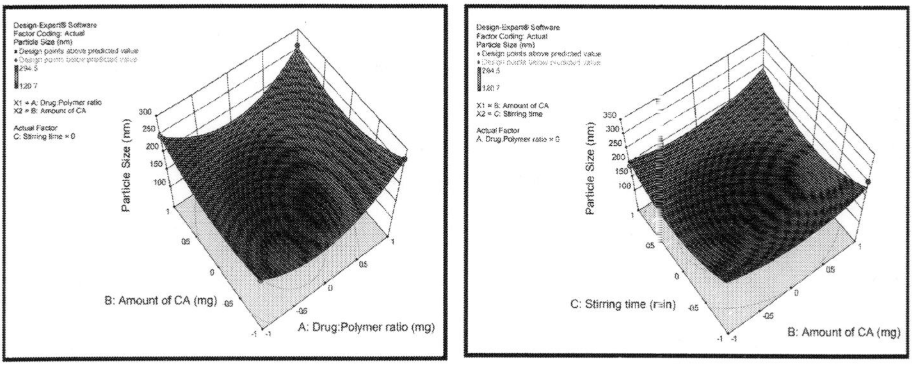

Figure 4 3D response surface curves for particle size

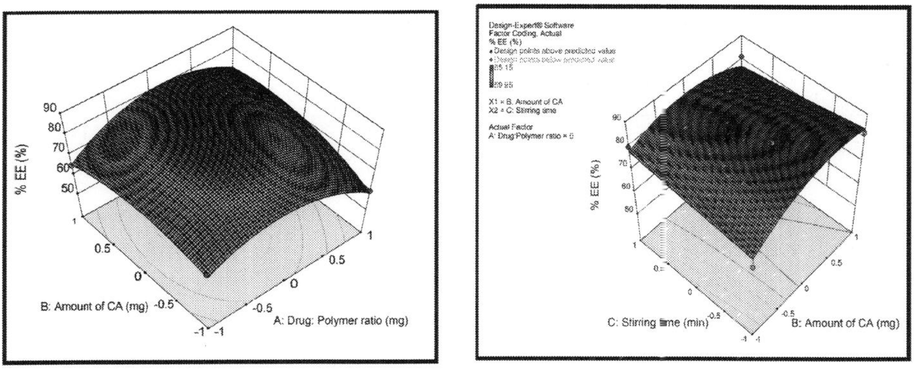

Figure 5 3D response surface curves for % EE.

The statistical validation of the polynomial equations was established by ANOVA provision available in the software (Table 5). Desirability function was utilized to optimize the best batch. The desirability of the optimized batch was found to be 0.978. The contour plot and 3D surface plot showing desirability of optimized batch are shown in Figure 6. The optimized batch was having particle size 121.2 ± 2.7 nm and 85.15 ± 1.1 %EE.

Zeta Potential

The zeta potential of optimized batch was found to be +26.6 mV which ensures that the CSNPs loaded with GCV were stable. Moreover, the sialic acid residues are present on mucin have pKa of 2.6, making them negatively charged at physiological pH. The prepared GCV loaded CSNPs had positive surface charge, which is favourable for the electrostatic interaction with the mucin of mucous and imparts mucoadhesive characteristics.26

Response	Source	DF[a]	SS[b]	MS[c]	F ratio[d]	p value
Y_1	Model	9	48816.51	5424.06	9.71	0.111
	Error	5	2794.25	558.85		
		Total	14	51610.76		
Y_2	Model	9	823.00	91.44	5.93	0.0321
	Error	5	77.10	15.42		
		Total	14	900.10		

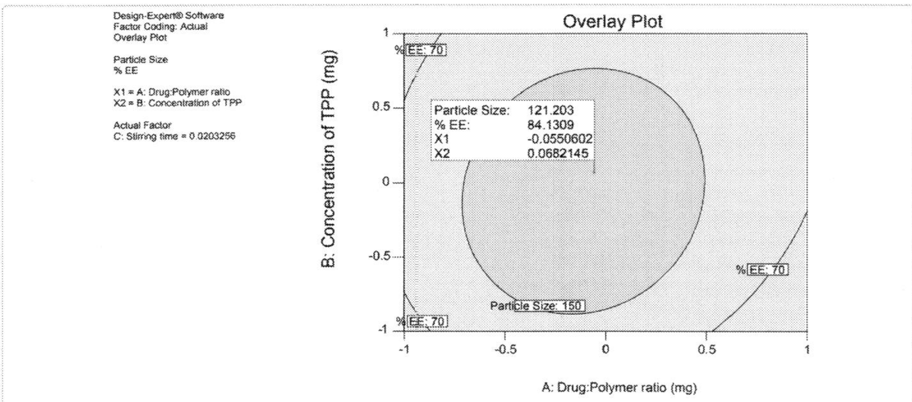

Figure 6 Overlay contour plot showing the location of the desirable region for selecton of optimized Nanoparticle formulation of GCV.

Morphology Observation

The TEM and SEM images (Figure 7) revealed that particles were spherical in shape and with the size which correlated with the dynamic light scattering particle size measurement.

FTIR spectroscopy

Figure 8 shows the FTIR spectra of GCV, CS and formulation. In the FTIR spectra of cross-linked CS the peak of 1655 cm–1 disappears and two new peaks at 1645 cm–1 and 1554 cm–1 appear. The disappearance of the band could be attributed to the linkage between the phosphoric and ammonium ions. The cross linked CS also showed a peak for P=O at 1155 cm–1 [25]. Cross-linked CS-TPP NPs also shows broader peak at 2934-3421 cm-1 indicates hydrogen bonding between OH and NH2 groups [26,27]. So, this confirms the crosslinking between CS and TPP.

Powder X-ray Diffraction (PXRD) Study

The XRD patterns are shown in Figure 9. The XRD patterns of CS show 19.78°, 29.29° 32.41°, 36.52° prominent crystalline peaks (2θ values) with high intensity [27]. The 2θ value of GCV at 10.7°, 14.9°, 20.0°, 23.4°, 23.8°, 25.1°, 26.1°, and 28.7° shows strongly indicates crystalline nature of GCV [28]. The powder diffraction profiles obtained for GCV corresponded with the XRD pattern reported by Ruchira Maiti Sarbajna [29]. The XRD diffractogram of prepared GCV loaded CSNPs suggested a lack of crystallinity or diffraction patterns were absent which indicats the existence of GCV in amorphous form or molecular dispersion [30]. XRD of CS shows few peaks, which indicates non-crystallinity. In GCV loaded CSNPs formulation a shift of peak positions, reduction of peak intensity and broadness of peaks were observed. It shows the destruction of the native CS packing structure due to the crystallized CS converted to amorphous form after cross-linked with TPP [31].

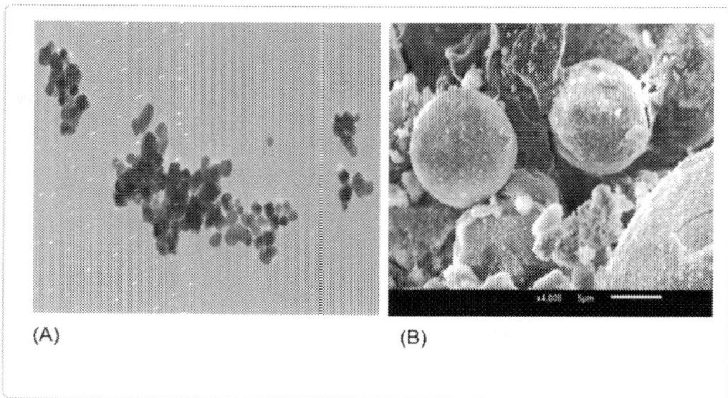

Figure 7 (A) TEM and (B) SEM micrograph of GCV loaded CSNPs.

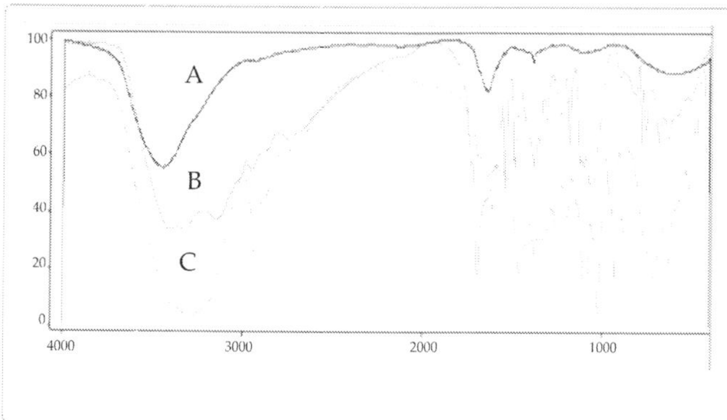

Figure 8 FTIR spectra of (A) Chitosan, (B) GCV (C) GCV loaded CSNPs.

Conductivity Study

The conductivity studies were performed to determine the relative changes in the ionic species with the change in pH. This would represent the nature of cross-linking with the change in the pH of TPP. The Figure 10 shows the conductivity curve. At first, when CS is solubilized in 1% glacial acetic acid, (pH 3.2) the free amino groups get hydrated. Addition of TPP at lower pH 3 leads to a decrease in conductivity up to 9 mL of TPP solution. When TPP at pH 3 containing Tripolyphosphate ($P3O10^{-5}$) ions was added to the CS solution; it interacts with the amino (–NH+3) group. This in turn leads to a decrease in the conductivity.

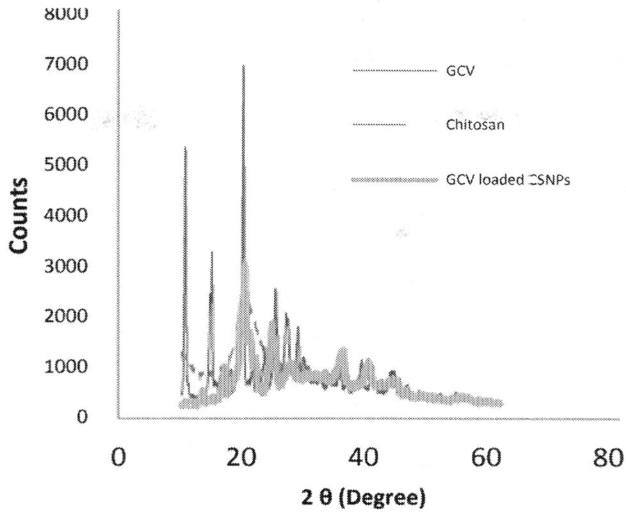

Figure 9 Overlay XRD pattern for chitosan, GCV, GCV loaded CSNPs.

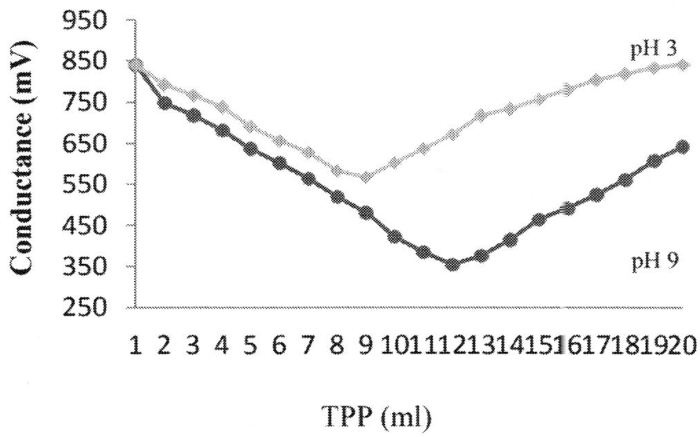

Figure 10 Effect of pH on conductivity during cross-linking of chitosan with TPP.

When a further amount of TPP is added, all the amino groups get saturated with P3O5^{-10} ions and the conductivity rises after 9 mL addition of TPP because of the increase in H+ ions. The process is predominantly ionic cross-linked. When TPP at higher pH (pH 9) was added to CS solution, the decrease in conductance was observed up to addition of 12 mL of TPP. As stated above, in TPP at higher pH (pH 9) both the OH– and phosphoric ions are present, which compete with each other to interact with the –NH+3 sites of CS. The OH– ions are linked to the amino groups by deprotonation [15].

In-vitro Drug Release Study

The *in-vitro* drug release study (Figure 11) was performed in pH 6.8 phosphate buffer. Drug exhibited sustained release behaviour with a steady rise in cumulative drug release up to the 24 hr. This shows uniform distribution of drug in the formulation. It was noteworthy that formulation showed an initial burst release possibly due to the small size of the NPs. As the particle diameter was reduced, the specific surface area increased, while the path length to the surface of the drug decreased.5 Various kinetic models with their R2 and SSR values are shown in Table 6. The model having highest value of R2 and lowest value of SSR is considered as the best model to fit [32].

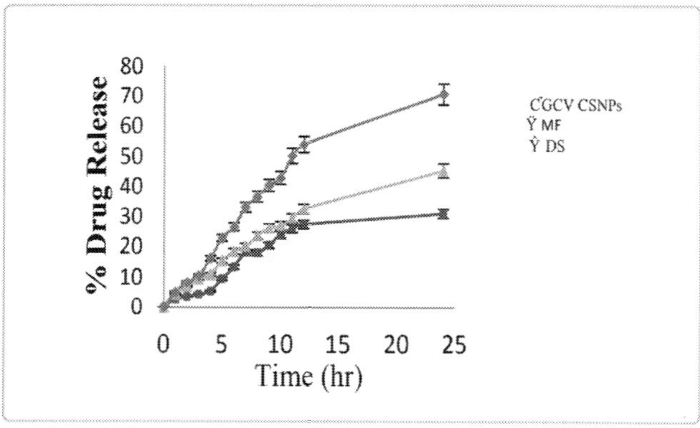

Figure 11 Release profile of GCV loaded CSNPs in phosphate buffer 6.8

Table 6 Correlatio coefficients and sum squared of residual (SSR) values of various kinetics models for drug release of GCV loaded CSNPs (Mean n=6 + SD).

Dissolution medium	Phosphate buffer pH 6.8			
Model	Zero order	First order	Higuchi	Hixon-Crowell
R^2 (Goodness of fit) values	0.8819	0.5929	0.9078	0.624
SSR values	41.30	196.55	0.3796	466.59

From the results it was found that drug release from CSNPs follows Higuchi model indicating that drug release as a diffusion process based on the Fick's law [24].

CONCLUSION

In the present study, GCV was entrapped in CSNPs formulation. After the identification and screening of the various formulation and process variables, the formulation was optimized. The optimum formulation had a relatively low particle size with a high %EE, a positive zeta potential, and optimal drug release profile, and is therefore potentially suitable for oral administration. The conductivity study, FTIR, and PXRD analysis confirmed the crosslinking mechanism. Enhanced GCV permeability was found in CSNPs loaded formulation in comparison with marketed formulation (P>0.05). The confocal microscopy resulted with effective penetration of GCV loaded CSNPs. Thus, incorporation of GCV into CSNPs results in enhanced permeability, that may in turn increase overall oral absorption of the drug. Further studies like cytotoxicity, cellular uptake through Caco-2 cell line and *in-vivo* animal study were carried out to verify the quality and performance of the CSNPs and will be included in the second part of the paper.

REFERENCES

1. Anderson RD, Griffy KG, Jung D, Door A, Hulse JD, et al. (1995) Ganciclovir absolute bioavailability and steady-state pharmacokinetics after oral administration of two 3000-mg/d dosing regimens in human immunodeficiency virus - and cytomegalovirus - seropositive patients. ClinTher 17: 425-32.
2. Andrei G, Snoeck, Robert N, Johan S, Marit LM, et al. (2000) Antiviral activity of ganciclovirelaidic acid ester against herpesviruses. AntiviralRes 45: 157-67.
3. Li M, Si L, Pan H, Rabba AK, Yan F, Qiu J, et al. (2011) Excipients enhance intestinal absorption of ganciclovir by P-gp inhibition: assessed in vitro by everted gut sac and in situ by improved intestinal perfusion. Int J Pharm 403: 37-45.

4. Yu CY, Cao H, Zhang XC, Zhou FZ, Cheng SX, et al.(2009) Hybrid nanospheres and vesicles based on pectin as drug carriers. Langmuir 25: 11720-26.
5. Cordova M, Cheng M, Trejo J, Johnson SJ, Willhite GP, et al. (2008) Delayed HPAM gelation via transient sequestration of chromium in polyelectrolyte complex nanoparticles. Macromol 41: 4398-404.
6. Ye YJ, Wang Y, Lou KY, Chen YZ, Chen R, et al. (2015) The preparation, characterization, and pharmacokinetic studies of chitosan nanoparticles loaded with paclitaxel/dimethyl-β-cyclodextrin inclusion complexes. Int J Nanomed 10: 4309.
7. Merodio M, Arnedo A, Renedo MJ, Irache JM (2001) Ganciclovir-loaded albumin nanoparticles: characterization and in vitro release properties. Eur J Pharm Sci 12: 251-59.
8. Ren J, Zou M, Gao P, Wang Y, Cheng G (2013) Tissue distribution of borneolmodified ganciclovir-loaded solid lipid nanoparticles in mice after intravenous administration. Eur J Pharma Biopharm 83: 141-48.
9. Nicolazzi Cl, Venard V, Le Faou A, Finance C (2002) In vitro antiviral efficacy of the ganciclovircomplexed with Î²-cyclodextrin on human cytomegalovirus clinical strains. Antiviral Res 54: 121-27.
10. Kajiwara E, Kawano K, Hattori Y, Fukushima M, Hayashi K, et al. (2007) Longcirculating liposome-encapsulated ganciclovir enhances the efficacy of HSVTK suicide gene therapy. J Control Rel 120: 104-10.
11. Nagavarma B, Yadav HK, Ayaz A, Vasudha L, Shivakumar H (2012) Different techniques for preparation of polymeric nanoparticles-a review. Asian J Pharm Clin Res 5: 16-23.
12. Plapied L, Duhem N, Des RA, Préat V (2011) Fate of polymeric nanocarriers for oral drug delivery. Curr Opin Colloid Interface Sci 16: 228-37.
13. Beloqui A, Solinis MA, Gascon AR, PozoRodríguez A, Rieux A, et al. (2013) Mechanism of transport of saquinavir-loaded nanostructured lipid carriers across the intestinal barrier. J Control Rel 166: 115-23.

14. Calvo P, Remunan LC, Vila JJ, Alonso M (1997) Novel hydrophilic chitosanpolyethylene oxide nanoparticles as protein carriers. J Appl Polym Sci 63: 125-32.
15. Calvo P, Remunan LC, Vila JJ, Alonso M (1997) Chitosan and chitosan/ethylene oxide-propylene oxide block copolymer nanoparticles as novel carriers for proteins and vaccines. Pharm Res 14: 1431-36.
16. Egashira K (2007) Pharmaceutical composition containing statin-encapsulated nanoparticle. US Patents 20100085602 A1 2007.
17. Kumari A, Yadav SK, Yadav SC (2010) Biodegradable polymeric nanoparticles based drug delivery systems. Colloids Surf B 75: 1-18.
18. Desai MP, Labhasetwar V, Amidon GL, Levy RJ (1996) gastrointestinal uptake of biodegradable microparticles: effect of particle size. Pharm Res 13: 1838-45.
19. Meng J, Sturgis TF, Youan B-BC (2011) Engineering tenofovir loaded chitosan nanoparticles to maximize microbicidemucoadhesion. EurJPharmSci 44: 57-67.
20. Elmowafy E, Osman R, El-Shamy AH, Awad G (2014) Stable Colloidal Chitosan/Alginate Nanocomplexes: Fabrication, Formulation Optimization and Repaglinide Loading. Int J Phar Pharm Sciences 6: 520-525.
21. Ray M, Pal K, Anis A, Banthia A (2010) Development and characterization of chitosan-based polymeric hydrogel membranes. Des Monomers Polym 13: 193-206.
22. Qi L, Xu Z, Jiang X, Hu C, Zou X (2004) Preparation and antibacterial activity of chitosan nanoparticles. Carbohydr Res 339: 2693-700.
23. Bhumkar DR, Pokharkar VB (2006) Studies on effect of pH on cross-linking of chitosan with sodium tripolyphosphate: a technical note. Aaps Pharmscitech 7: E138-E43.
24. Dash S, Murthy PN, Nath L, Chowdhury P (2010) Kinetic modeling on drug release from controlled drug delivery systems. Acta Pol Pharm 67: 217-23.
25. Dangi R, Shakya S (2013) Preparation, optimization and characterization of PLGA nanoparticle. Int J Pharm Life Sci.

26. Honary S, Zahir F (2013) Effect of zeta potential on the properties of nano-drug delivery systems-a review (Part 1). Trop J Pharm Res 12: 255-64.
27. Anbarassan B, Vennya V, Niranjana V, Ramaprabhu S (2013) Optimization of The Formulation and In-vitro Evaluation of Chloroquine Loaded Chitosan Nanoparticles Using Ionic Gelation Method. J Chem Pharm Sci 6: 66-72.
28. Joshi S, Bilgaiyan P, Pathak A (2012) Synthesis and in-vitro study of some medicinally important mannich bases derived from 2-amino-9 [{(1, 3 dihydroxy propane-2yl) oxy} methyl] 6-9 dihydro-3h-purin-6-one. J Chil Chem Soc 57: 1277-82.
29. Sarbajna RM, Preetam A, Devi AS, Suryanarayana M, Sethi M, et al. (2011) Studies on Crystal Modifications of Ganciclovir. Mol Cryst Liq Cryst 537: 141-54.
30. Guo P, Hsu TM, Zhao Y, Martin CR, Zare RN (2013) Preparing amorphous hydrophobic drug nanoparticles by nanoporous membrane extrusion. Nanomed 8: 333-341.
31. Vaezifar S, Razavi S, Golozar MA, Karbasi S, Morshed M, et al. (2013) Effects of some parameters on particle size distribution of chitosan nanoparticle prepared by ionic gelation method. J Cluster Sci 24: 891-903.
32. Singh J, Gupta S, Kaur H (2011) Prediction of in vitro Drug Release Mechanisms from Extended Release Matrix Tablets using SSR/R^ sup 2^ Technique. Trends Appl Sci Res 6: 400.

10
CHAPTER

GREEN SYNTHESIS OF COPPER-CHITOSAN NANOPARTICLES AND STUDY OF ITS ANTIBACTERIAL ACTIVITY

Appu Manikandan and Muthukrishnan Sathiyabama

Department of Plant Science, Bharathidasan University, Tiruchirappalli, Tamil Nadu, India

ABSTRACT

Copper nanoparticle (Cu-NP) synthesis has been gaining attention due to its property and applicability. The main objective of the work is to synthesize stable Cu-NP through green synthesis. This work demonstrates the synthesis of Cu-NP by the addition of the acidic chitosan solution to CuSO4 solution with constant stirring at 70°C for 12 h. The addition of chitosan aids the stable formation of nanoparticle. The formation of nanoparticle was observed by the peak at 536 nm using UV-Vis spectroscopy. The interaction between the chitosan and the synthesized nanoparticles was studied using FTIR spectroscopy. The average size of the nanoparticle as determined by zetasizer was about 88.21 nm, while the zeta potential was -29.0 mV. Transmission electron microscopy (TEM) analysis indicated the spherical size with a size ranging between 20-30 nm. The Energy-dispersive X-ray spectroscopy (EDS) analysis showed a homogenous copper-rich composition of metal nanoparticles. The X-ray diffraction (XRD) analysis confirmed the presence of copper nanoparticle with crystalline nature. The synthesised Cu-chitosan nanoparticle exhibited antibacterial activity against gram negative as well as gram positive bacteria. However, the activity was more against gram negative bacteria which may be due to the difference in cell wall composition.

INTRODUCTION

Metal NPs have attracted considerable attention as an antimicrobial agent due to their high surface area to volume ratio, which allows NPs to be effective in very small amounts [1]. An increasing interest in Cu-NPs arises from the useful properties of this metal (thermal and electrical conductivity), which is achieved at much lower cost compared to other noble metals like silver and gold [2]. Cu is known to have antibacterial and antifungal properties [3,4] also it is non-toxic to mammals [5].

These properties thus make Cu-NP synthesis an attractive area. A number of methods for producing Cu-NPs have been developed using both physical and chemical approaches, which involves elevated temperatures, inert atmospheres, large amount of surfactants and organic solvents [6-10]. However, the major limitations in the synthesis of CuNPs are their ease of oxidation [11] to CuO or Cu2O during preparation and storage [12]. Therefore, alternative methods have been developed to synthesize CuNP in the presence of polymers and surfactants as stabilizer [11,13]. There are some recent reports on a successful synthesis of Cu-NPs in aqueous solution, using chitosan which involve additional reducing and stabilizing agents at various steps [14-18] that are associated with environmental toxicity [19].

Thus, there is still a need for the development of a method for synthesis of stable Cu-NPs, without involving much co-reducing/ stabilizing agents for antimicrobial application.

This work is an attempt to propose a novel one step green synthesis of Cu-NPs with colloidal stability in aqueous media using chitosan, a biopolymer as a reducing and capping agent. The synthesis of Ag-NPs using chitosan as both a reducing and capping agent has been reported [20]. The choice of Chitosan as a stabilizer of the Cu-NPs is because of its ability to chelate metals, which makes a perfect material for metal NP synthesis [21]. Synthesis of NPs using polymer has been promising due to their ease of processing, solubility, biocompatibility, less toxicity and also because of the possibility of controlling the growth of the resulting NPs. Travan et al. [22] demonstrated that hydrophilic side chains of chitosan play a fundamental role in stabilization and dispersion of NPs, preventing their agglomeration. Chitosan can co-ordinate metal ions (eg. Cu ions) before reduction. When Cu salts dissolve in acidified Chitosan solution, Cu ion binds to the polymer chains via amino groups. The reduction of these ions takes place further, coupled with the oxidation of the hydroxyl groups [23,24]. Thus NPs generated are strongly attached to the chitosan due to the chemical bond between the electron rich nitrogen present in the amino groups of the polymer and copper [25,26]. The use of biopolymer as capping and or reducing agents represents an environmentally friendly alternative to hazardous organic solvents [27].

In the present study, we report the environmentally friendly synthesis of stable Cu chitosan NPs through green route with chitosan as a stabilizer and or reducing agent. The synthesized NPs were characterized by spectrometry, zetasizer, TEM, XRD and EDS. Their antimicrobial properties were tested using gram negative and gram positive bacteria.

MATERIALS AND METHODS

Synthesis of Cu-Chitosan NPs

In a typical one-step synthesis protocol 50 ml of 0.75% (w/v) chitosan solution was prepared using 0.1% acetic acid (in distilled water) solution and 25 ml of 50 mM CuSo4. 5H2O was then added to it and stirred on magnetic stirrer at 70°C for 12 h till the reaction was completed. After 12 h the colloid was centrifuged at 10,000 G for 10 mins to separate particles from suspension. The precipitate was re suspended in acetone (90%, v/v) and the centrifugation was repeated three times to remove unreacted reagents. Finally, the precipitate was dissolved in water, dried under vacuum overnight and stored.

Characterization of Cu-Chitosan NPs

UV-visible spectra were recorded using a Shimadzu UV-Vis1800 Spectrophotometer for the confirmation of NP formation. The structural features of Cu-Chitosan NPs were performed in a Nicolet 560 FTIR spectrometer in a range from 400 to 4000 cm-1 using a KBr pellet technique. Particle size analysis was performed by dynamic light scattering (DLS). The charge on the surface of the particles was characterized by measuring the zeta potential of the suspension using

a Zetasizer (Malvern Instruments, UK). The size and morphology of the Cu-Chitosan NPs were examined by HRTEM (JEOL model 1200 EX). EDS (Energy- dispersive X-ray spectroscopy, Bruker) was applied to study the purity and chemical composition of the samples. X-ray diffraction studies were performed with an X-ray diffractometer (Rigaku Ultima III XRD) with Cu $K\alpha 1$ radiation to determine the structure of a sample. The X-ray source was operated at 40 kV and 40 mA. Diffraction intensity was measured in the reflection mode at a scanning rate of 2°/min for 2θ=10-70°.

Determination of Antimicrobial Activity

A disk diffusion method was used to assay the antibacterial activity of synthesized Cu-Chitosan NPs (50 µg/ml), against test strains in Muller-Hinton agar media. For control experiments, ampicillin and streptomycin were used for gram negative (*E.coli, Salmonella paratyphi*) and gram positive bacteria (*Bacillus*) respectively. Chitosan was also used separately to check the antibacterial activity. The plates were incubated for 24 h at 37°C and inhibition zones were measured.

RESULTS AND DISCUSSION

The chemical reduction of metal salt is one of the most convenient and promising synthetic approaches to obtain metallic nanoparticles with relatively inexpensive setups. Cu-NPs are very difficult to obtain by simple reduction of copper salts in aqueous solution, when compared with other metals such as Ag, Au or Pt, since copper oxidizes to CuO and Cu2O [25,28,29]. However, the size controlled synthesis of Cu-NPs can be obtained by using a carrier material, such as a polymer matrix e.g., chitosan which is having a superior copper binding properties [17,29].

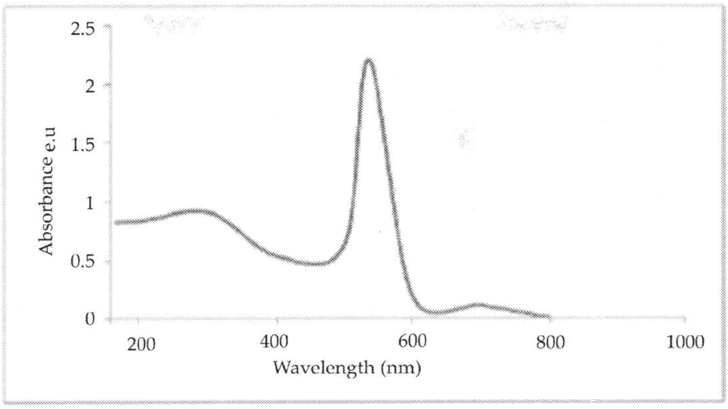

Figure 1 UV-visible spectrum of synthesized copper-chitosan nanoparticle.

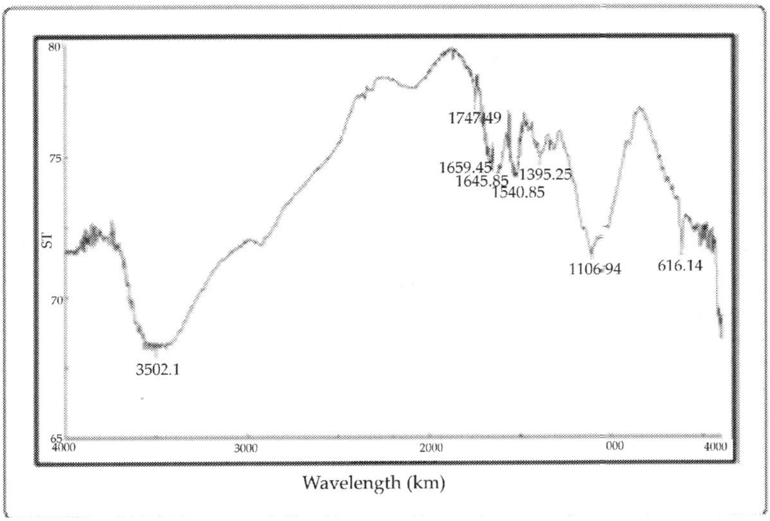

Figure 2 FT-IR spectra of Cu-Chitosan nanoparticle.

This polymer has been applied as a reducing agent and or a stabilizing agent in the formation of Cu-NPs, which inhibits agglomeration, minimizes oxidation reaction thus limits the particle growth. The evolution of the absorption spectra of different aliquots overtime of a solution containing copper sulphate (50 mM) in chitosan (0.75% w/v) showed major peak around 500-600 nm (data not shown), thus revealing the formation of Cu-NPs. The colloidal Cu-NPs prepared at 12 h at 70°C showed a characteristic single absorption peak at 536 nm (Figure 1), which indicates the reduction of Cu2+ ions in the presence of chitosan to highly monodisperse NPs. There are many reports on the blue shift of the surface plasmon resonance (SPR) of the metallic NPs with their decreasing size [30]. Nanosized Cu particles generally exhibit an SPR at around 500-600 nm [31]. When copper sulphate was added to the chitosan solution, Cu2+ ions would attach to chitosan macromolecules by electrostatic interactions. The electron-rich oxygen atoms of polar hydroxyl and ether groups of chitosan are likely to interact with electropositive metal cations [17]. It has been reported that chitosan acts as a controller of nucleation as well as a stabilizer [32].

FTIR analysis was conducted to determine the molecular interactions between chitosan and the synthesized NPs. FTIR analysis shows the presence of bands at 3502.1 cm-1 (due to overlap of O-H and N-H stretching), 1747.19 cm-1 (C=O stretching), 1659.45,1645.85 and 1540.85 cm-1 (NH2 bending), 1395.25 cm-1 (C-H bending), 1106.94 cm-1 (C-O-C stretching) 616.14 cm-1 (Figure 2). The peak at 616.14 cm-1 evidence the interaction between Cu-NPs and chitosan, which indicates that NPs were capped by the polymer [12,13]. The experiments described above indicate the importance of chitosan in the performed synthesis. Chitosan, as a biopolymer, has a significant content of primary amines and hydroxyl groups and possess a strong affinity towards metal ions, which are incorporated by simple chelation by ion exchange, thereby making it an excellent support for nucleation of Cu-NPs.

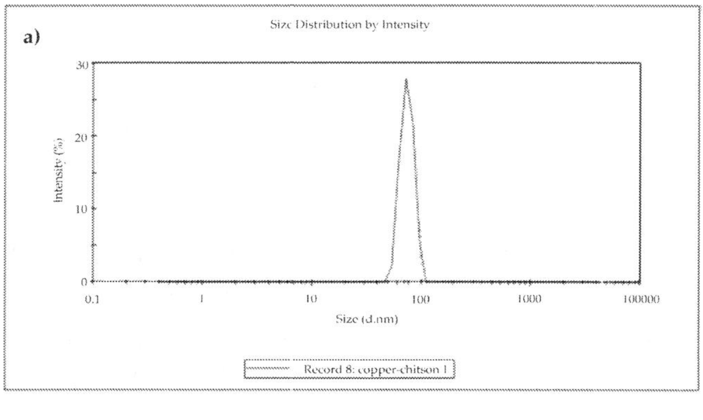

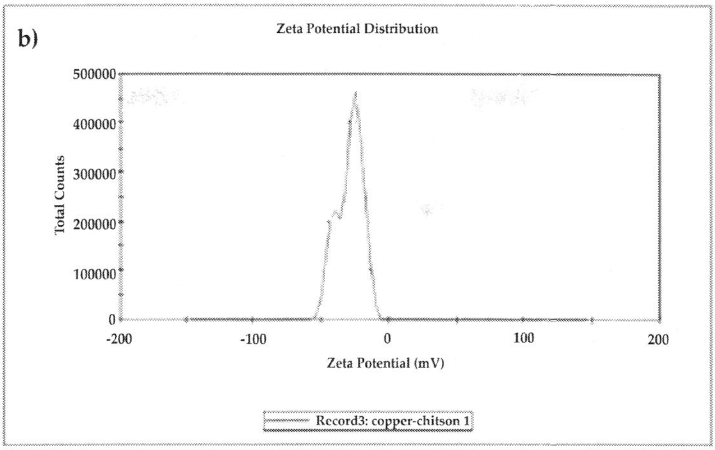

Figure 3 Dynamic light scattering (a) and zeta-potenitial of Cu-chitosan nanoparticle (b).

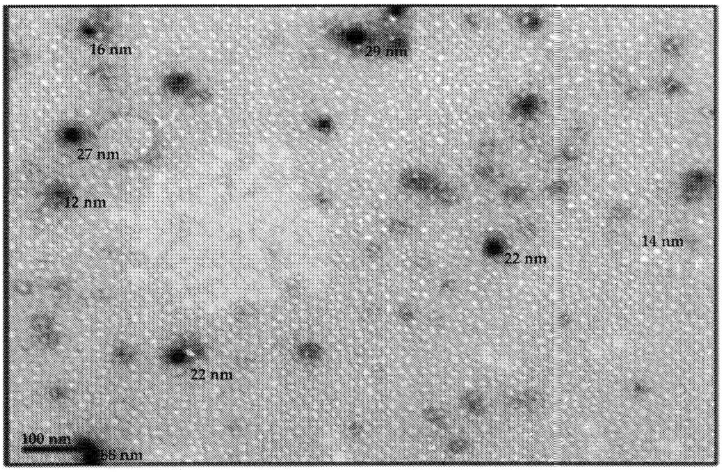

Figure 4 HRTEM micrograph of synthesized copper-chitosan nanoparticle scale bar =100 nm.

The generation of Cu2+ Chitosan complexes in solution enables metal ion reduction. From these results, it can be envisaged that the biopolymer plays an important role as polymeric capping agent (Stabilizer). There are reports that chitosan plays a more specific role as a stabilizing agent [25].

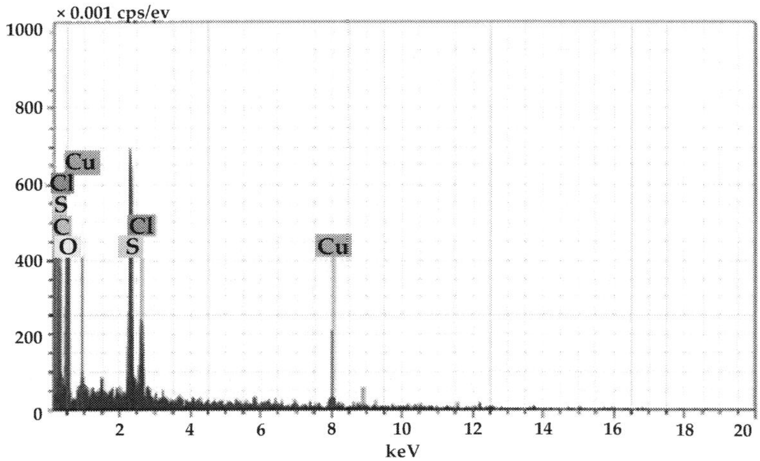

Figure 5 Energy-dispersive spectroscopy spectrum of cu-chitosan nanoparticle. Various X-ray emission peaks are labelled.

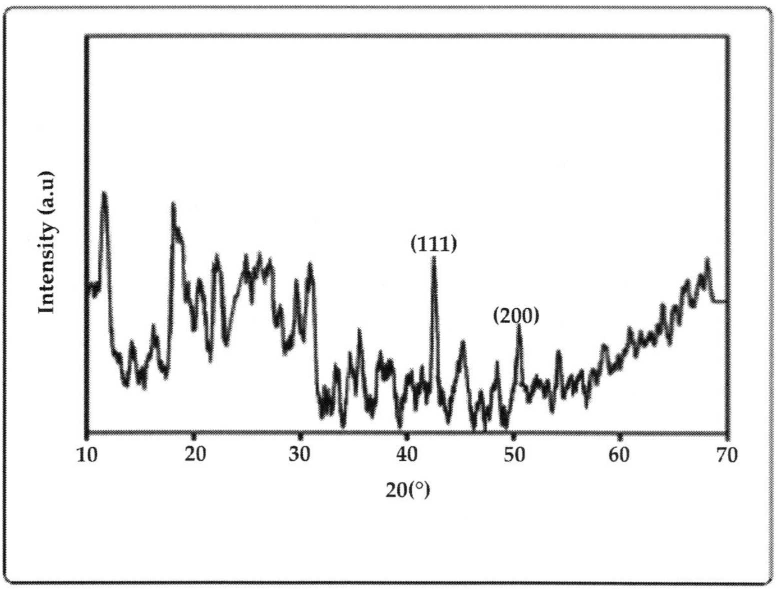

Figure 6 X-ray diffraction patterns of synthesized copper-chitosan nanopar-

ticles.

The average size of Cu-chitosan NP measured by zetasizer was 88.21 nm (Figure 3a) with a narrow size distribution (PDI-0.45). Zeta potential is an important parameter for determining the stability of nanoparticle suspensions. For a physically stable NP suspension to be stabilized solely by electrostatic repulsion, a zeta potential of ±30 mV is required as a minimum [31]. The mutual repulsion of NPs depends on having either a large negative or positive zeta potential. Zeta potential measurements of Cu-Chitosan NPs are presented in (Figure 3b) which shows that the synthesized NPs have a negative surface charge (-29.0 mV). This implies the colloidal stability of the nanoparticles. Chitosan can form various chemical bonds with metal components, thus enhancing the stability of the NPs [32] and prevents agglomeration.

Figure 4 shows a high-resolution electron microscope (HRTEM) micrograph of Cu-NPs which are spherical in shape and size range of about 20-30 nm. Further, TEM images of the sample corroborated the stability of the NPs in the presence of chitosan as there was no agglomeration and oxidation (Figure 4). The size difference between TEM and DLS analyses may be due to different principle involved in these two techniques. The DLS measurement involves hydrodynamic state of nanoparticle where as it is a dry state in TEM measurement [33,34]. EDS analysis showed a very homogenous copper-rich composition for metal NPs (Figure 5). The weak carbon and oxygen signals may be due to chitosan that surrounds the Cu-NPs [25,35].

X-ray diffraction patterns for the Cu-Chitosan nanoparticles, was shown in The diffraction angles at 43.02°, 50.62°, which correspond to the characteristic face centered cubic (fcc) of copper lines indexe at (111), (200), respectively [36] indicates the crystalline nature. The diffraction angle observed at 20.54 is related to the chitosan [37] present in the nanoparticle. Further the presence CuO or Cu2O was not detected in the diffraction.

Antimicrobial Activity

The Chitosan stabilized Cu-NPs exhibited inhibitory activity towards gram negative, positive bacteria (Figure 7a). However, more antibacterial activity was observed against gram negative bacteria (Figure7b).

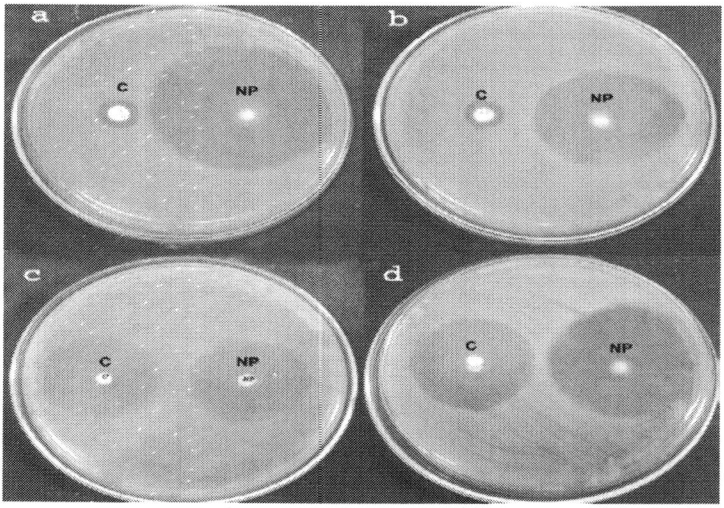

Figure 7a: Antimicrobial activity of Cu-Chitosan nanoparticle against gram negative and gra, positive bacteria by disk diffusion method. a) E.coli, b) Salmonella, c) Staphylococcus aureus, d) Bacillus.

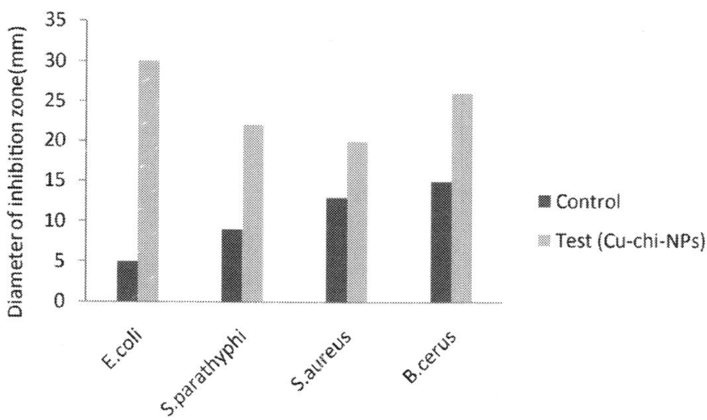

Figure 7b Effect of Cu-Chitosan nanoparticle on inhibition (in mm) of growth of gram negative and gram positive bacteria on Muller-Hinton Agar plate.

This may be due to the differences in the bacterial cell wall structure of gram negative (single peptidoglycan layer) to the gram positive organism (several peptidoglycan layers). The adhesion of NPs to the surface of a bacterium alters its membrane properties ultimately causing death [38]. The results of the present study are in agreement with other reports that indicate greater activity in copper NP against gram negative microorganism [39]. Further chitosan demonstrated only negligible zone of inhibition against test bacteria (data not shown) suggesting a lack of antimicrobial activity of chitosan at pH 7.4, which is the pH of the Muller-Hinton agar. This is in accord with the fact that chitosan has antimicrobial activity only in an acidic medium due to its poor solubility above pH 6.5 [40].

CONCLUSION

Colloidal stable Cu-chitosan NPs with a size range of 20-30 nm have been successfully synthesized through green synthesis in the presence of a biopolymer, chitosan. The synthesized NPs were characterized by UV-Vis, FTIR, TEM, EDS and XRD analyses. The main role attributable to the Chitosan is related to its chelating capacity. The present singlestep synthetic method represents a simple, convenient, cost-effective and eco-friendly alternate to other synthetic methods. Our results indicate the potential of copper NPs for combating pathogenic microorganisms. This may be suitable for formulation of new types of antimicrobial materials for pharmaceutical and biomedical application. However, further *in vivo* studies to determine the toxicity of these NPs are necessary.

REFERENCES

1. Sundaresan K, Sivakumar A, Vigneswaran C, Ramachandran T (2012) Influence of nano titanium dioxide finish, prepared by sol-gel technique, on the ultraviolet protection, antimicrobial, and self-cleaning characteristics of cotton fabrics. J Indust Text 41: 259-277.
2. Han WK, Choi JW, Hwang GH, Hong SJ, Lee JS, et al. (2006) Fabrication of Cu nano particles by direct electrochemical reduction from CuO nano particles. Appl Surface Sci 252: 2832-2838.

3. Hsiao SH, Fang TH, Hwang JS (2006) The bioconcentration of trace metals in dominant copepod species off the northern Taiwan coast. Crustaceana 79: 459-474.
4. Iakovidis I, Delimaris I, Piperakis SM (2011) Copper and Its Complexes in Medicine: A Biochemical Approach. Mol Bio Int.
5. Fleming CA, Trevors JT (1989) Copper toxicity and chemistry Indeed, in the environment: A review. Water Air Soil Pollut 44: 143-158.
6. Arul DN, Paul RC, Gedanken A (1998) Synthesis, characterization, and properties of metallic copper nanoparticles. Chem Mater 10: 1446-1452.
7. Joshi S, Patil S, Iyer V, Mahumuni S (1998) Radiation induced synthesis and characterization of copper nanoparticles. Nanostruct Mater 10: 1135-1144.
8. Cheng X, Zhang X, Yin H, Wang A, Xu Y, et al. (2006) Modifier effects on chemical reduction synthesis of nanostructured copper. Appl Surf Sci 253: 2727-2732.
9. El-Nour KM, Eftaiha A, Al-Warthan A, Ammar RA (2010) Synthesis and applications of silver nanoparticles. J Chemist 3: 135-140.
10. Thakkar KN, Mhatre SS, Parikh RY (2010) Biological synthesis of metallic nanoparticles. Nanomed Nanotech Bio Med 6: 257-262.
11. Wu SH, Chen DH (2004) Synthesis of high-concentration Cu nanoparticles in aqueous CTAB solutions. J Colloid Interface Sci 273: 165-169.
12. Usman MS, ezzat el M, Shameli ZK, Salama NZ, Ibrahim NA, et al. (2013) Synthesis, characterization, and antimicrobial properties of copper nanoparticles. I J Nanomed 17: 4467-4479.
13. Dang TM, Le TT, Fribourg-Blanc E, Dang MC (2011) The influence of solvents and surfactants on the preparation of copper nanoparticles by a chemical reduction method. Adv Nat Sci Nanosci Nanotechnol 2.
14. Wu C, Mosher BP, Zeng T (2006) One-step green route to narrowly dispersed copper nanocrystals. J Nanopart 8: 965-969.
15. Xiong J, Wang Y, Xue Q, Wu X (2011) Synthesis of highly stable dispersions of nanosized copper particles using L-ascorbic acid.

Green Chem 13: 900-904.

16. Saharan V, Mehrotra A, Khatik R, Rawal P, Sharma SS, et al. (2013) Synthesis of chitosan based nanoparticles and their in vitro evolution against phytopathogenic fungi. I J Biol Macromol 62: 677-683.
17. Zain NM, Stapley AG, Shama G (2014) Green synthesis of silver and copper nanoparticles using ascorbic acid and chitosan for antimicrobial applications. Carbohydr Polym 112: 195-202.
18. Tokarek TW, Huo JA, Odame-Ankrah CA, Hammoud D, Taha YM, et al. (2014) A gas chromatograph for quantification of peroxycarboxylic nitric anhydrides calibrated by thermal dissociation cavity ring-down spectroscopy. Meas Tech Discuss 7: 5953-6019.
19. Lim S, Hudson SM (2004) Synthesis and antimicrobial activity of a water soluble chitosan derivative with a fiber reactive group. Carbohydr Resear 339: 313-319.
20. Sanpui P, Murugadoss A, Durgaprasad PV, Ghosh SS, Chattopadhyay A, et al. (2008) The antibacterial properties of a novel chitosan-Ag-nanoparticle composite. Int J Food Microbiol 124: 142-146.
21. Hardy JJ, Hubert S, Macquarrie DJ, Wilson AJ (2004) Chitosan-based heterogeneous catalysts for suzuki and heck reactions. Green Chem 6: 53-56.
22. Travan A, Pelillo C, Donati I, Marsich E, Benincasa M, et al. (2009) Non-cytotoxic Silver Nanoparticle-Polysaccharide Nanocomposites with Antimicrobial Activity. Biomacromolecules 10: 1429-35.
23. Tran HV, Tran LD, Ba CD, Vu HD, Nguyen TN, et al. (2010) Synthesis, characterization, antibacterial and antiproliferative activities of monodisperse chitosan-based silver nanoparticles. Colloids Surf .A 360: 32-40
24. Wei D , Qian W (2008) Facile synthesis of Ag and Au nanoparticles utilizing chitosan as a mediator agent. Colloids Surf B. 62: 136-42
25. Tokarek K, Hueso JL, Kustrowski P, Stochel G (2013) Green Synthesis of Chitosan-Stabilized Copper Nanoparticles. E J Inor Che 28: 4940-4947.
26. Regiel A, Irusta S, Kyziol A, Arruebo M, Santamaria (2013) Preparation and characterization of chitosan - silver nanocomposite

films and their antibacterial activity against *Staphylococcus aureus*. Nanotechnology doi: 10.1088/0957- 4484/24/1//015101.

27. Pakiari AH, Jamshidi Z (2007) Interaction of amino acids with gold and silver clusters. J. Phys. Chem. A. 111: 4391-6
28. Mallick S, Sharma S, Banerjee M, Ghosh SS, Chattopadhyay A, et al. (2012) Iodine-Stabilized Cu Nanoparticle Chitosan Composite for Antibacterial Applications. ACS Appl Mater Interfaces 4: 1313-1323.
29. Mallick K, Witcomb MJ, Scurrell MS (2006) In situ synthesis of copper nanoparticles and poly (o-toluidine): a metal-polymer composite material. Eur Polym J 42: 670-675.
30. Huang H, Yuan Q, Yang X (2004) Preparation and characterization of metalchitosan nanocomposites. Colloids and Surfaces B: Biointerfaces 39: 31-37.
31. Singare DS, Marella S, Gowthamrajan K, Kulkarni GT, Vooturi R, et al. (2010) Optimization of formulation and process variable of nanosuspension: An industrial perspective. Int J Pharma 402: 213-220.
32. Muzzarelli RA (2011) Potential of chitin/chitosan-bearingmaterials foruranium recovery: An interdisciplinary review. Carbohydr Polym 84: 54-63.
33. Wu Y, Yang W, Wang C, Hu J, Fu S, et al.(2005) Chitosan nanoparticles as a novel delivery system for ammonium glycyrrhizinate. Int J Pharm 295: 235-245.
34. Khlebtsov BN, Khlebtsov NG (2011) On the Measurement of Gold Nanoparticle Sizes by the Dynamic Light Scattering Method. Colloid Journal 73: 118-127.
35. Jayaseelan C, Ramkumar R, Rahuman AA, Perumal P (2013) Green synthesis of gold nanoparticles using seed aqueous extract of Abelmoschus esculentus and its antifungal activity.Indust. Crops Produc 45: 423-429.
36. Mott D, Galkowski J, Wang LY, Luo J, Zhong CJ, et al.(2007) Synthesis of sizecontrolled and shaped copper nanoparticles. Langmuir 23: 5740-5745.
37. Ahmad MB, Lim JJ, Tay MY, Shameli K, Ibrahim NA, et al. (2011)

Synthesis of silver nanoparticles in chitosan, gelatin and chitosan/ gelatin bionanocomposites by a chemical reducing agent and their characterizations. Molecules 16: 7237- 7248.

38. Li Q, Mahendra S, Lyon DY, Brunet L, Liga M, et al. (2008) Antimicrobial nanomaterials for water disinfection and microbial control: Potential applications and implications. Water Research 42: 4591-4602.

39. Longano D, Ditaranto N, Sabbatini L, Torsi L, Cioffi N, etal. (2012) Nano- Antimicro Progress and Prospects 3: 85-117.

40. Rabea EI, Badawy ME, Stevens CV, Smagghe G, Steurbaut W, et al.(2003) Chitosan as antimicrobial agent: applications and mode of action. Biomacromolecule 4: 1457-1465.

11

CHAPTER

NANOPARTICLES BASED ON CHITOSAN AS CARRIERS FOR THE COMBINED HERBICIDES IMAZAPIC AND IMAZAPYR

Cintia Rodrigues Maruyama[1,2,3], Mariana Guilger[1,3], Mônica Pascoli[1,2,3], Natalia Bileshy-José[1], P.C. Abhilash[4], Leonardo Fernandes Fraceto[2] & Renata de Lima[1,3]

[1]Laboratory of Biotechnology, University of Sorocaba, Rodovia Raposo Tavares, km 92, 18023-000, Sorocaba, Brazil.

[2]Department of Environmental Engineering, São Paulo State University (UNESP), Sorocaba, Brazil.

[3] Federal University of São Carlos, Sorocaba campus, Rodovia João Leme dos Santos, km 110, 18052-780, Sorocaba, Brazil.

[4]Institute of Environment & Sustainable Development, Banaras Hindu University, Varanasi, India.

ABSTRACT

The use of lower concentrations and fewer applications of herbicides is one of the prime objectives of the sustainable agriculture as it decreases the toxicity to non-targeted organisms and the risk of wider environmental contamination. In the present work, nanoparticles were developed for encapsulation of the herbicides imazapic and imazapyr. Alginate/chitosan and chitosan/tripolyphosphate nanoparticles were manufactured, and their physicochemical stability was evaluated. Determinations were made of the encapsulation efficiency and release kinetics, and the toxicity of the nanoparticles was evaluated using cytotoxicity and genotoxicity assays. The effects of herbicides and herbicide-loaded nanoparticles on soil microorganisms were studied in detail using real-time polymerase chain reactions. The nanoparticles showed an average size of 400 nm and remained stable during 30 days of storage at ambient temperature. Satisfactory encapsulation efficiencies of between 50 and 70% were achieved for both types of particles. Cytotoxicity assays showed that the encapsulated herbicides were less toxic, compared to the free compounds, and genotoxicity was decreased. Analyses of soil microbiota revealed changes in the bacteria of the soils exposed to the different treatments. Our study proves that encapsulation of the herbicides improved their mode of action and reduced their toxicity, indicating their suitability for use in future practical applications.

INTRODUCTION

The use of chemicals to combat pests and diseases in agriculture is not a recent phenomenon. Substances such as sulfur, arsenic, lime, and nicotine have traditionally been employed for this purpose, although the post-industrial development of the chemical industry has greatly expanded the available range of products used as pesticides. However, inappropriate use of these substances can result in adverse environmental effects1. The herbicides imazapic and imazapyr are the members of the imidazolinone group of compounds mainly used to control weeds in plantations of maize, soybean, and peanut2. These compounds can control a broad spectrum of weeds, with effects at extremely low dosages, but a drawback is their high persistence in the soil3. The mechanism of action is based on inhibition of the acetolactate synthase enzyme responsible for catalysis of the first step in the synthesis of amino acids such as valine, leucine, and isoleucine, hindering the cellular development of the plant. The compounds are easily transported through the roots and stems. However, the development of resistance by plants has led to the combined use of imazapic and imazapyr, which has been successfully employed in cultivations of maize4.

In recent years, there has been an increased interest in nanotechnology as a means of improving the effectiveness of herbicides, while at the same time minimizing their environmental impacts5–10. Several studies have described the use of herbicide carrier systems that provide better control of weeds, without harming nontarget organisms8–10. Among the various nanostructured systems that can be used with herbicides, some of the most promising are based on polymeric substances11–16,8,17,9,10. One polymer that has been highlighted for use in the agricultural area is chitosan, a biodegradable substance obtained from the deacetylation of chitin, which is highly effective as a carrier system for agrochemicals and plant micronutrients18. Silva *et al.* (2011) developed a system based on nanoparticles of chitosan and alginate as carriers for the herbicide paraquat. The nanoparticles showed an average diameter of 635 nm, polydispersity index of 0.518, zeta potential of − 22.8 mV, and encapsulation efficiency of 74.2%. Encapsulation of the herbicide slowed its release and the results confirmed the effectiveness of using alginate in nanoparticles of chitosan14.

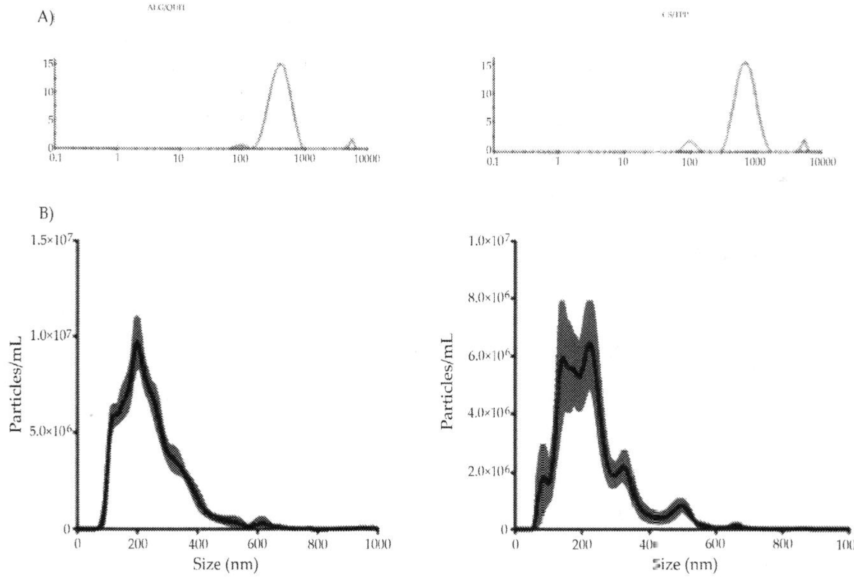

Figure 1. Chemical analysis of the CS/ALG and CS/TPP nanoparticles: (A) initial size distributions of the CS/ALG/IMC+IMR and CS/TPP/IMC+IMR formulations; (B) number concentrations of the CS/ALG/IMC+IMR and CS/TPP/IMC+IMR nanoparticles as a function of size.

Grillo *et al.* (2014a) prepared chitosan/tripolyphosphate nanoparticles as a carrier system for paraquat, and found that the encapsulated herbicide was as effective as the free form, but with a slower release profile. The particles showed an average size of 300 nm, encapsulation efficiency of 62%, and low genotoxicity8. Given this background, the aim of the present work was to prepare and characterize nanoparticles based on chitosan (chitosan/alginate – CS/ALG and chitosan/tripolyphosphate – CS/TPP), containing imazapir and imazapyc for use in agricultural weed control. There have been no previous reports of the combination of these herbicides in nanoparticles; therefore the development of effective encapsulation technique represents an important technological advancement in this area. The formulations were characterized in terms of size distribution, zeta potential, and polydispersity index, as well as their ability to encapsulate the active agents. Comparison was made between the nanoparticle systems and the free herbicides in terms of their cytotoxicity and genotoxicity, in order to confirm that the encapsulation procedure resulted in a reduction in toxicity. The effects of these systems on bacteria associated with the soil nitrogen cycle were evaluated in order to determine the likelihood of any potential impacts on soil microbiota. Finally, tests using a target species (*Bidens pilosa*) was conducted to determine the effectiveness of the encapsulated herbicides, with a view to their use as a safer alternative in agriculture.

RESULTS AND DISCUSSION

Characterization and Stability of The Formulations

Two different methodologies were used to determine the size distributions of the particles containing the herbicides: (i) dynamic light scattering and (ii) nanoparticle tracking analysis. The average size distributions of the NP-CS/TPP using the DLS and NTA methods were 478.6} 52.3 nm and 233.5} 10.7 nm, respectively (Fig. 1). The concentration of the CS/TPP nanoparticles containing the herbicides, calculated from the NTA results was 1.20 × 109 particles/mL. The average size distributions of the CS/ALG nanoparticles using the DLS and NTA methods were 377.7} 9.7 nm and 246.8} 2.6 nm, respectively (Fig. 1). The CS/ALG nanoparticles showed a higher concentration (1.83 × 109 particles/mL) and a size distribution that was more homogeneous (lower polydispersity), compared to the CS/TPP nanoparticles.

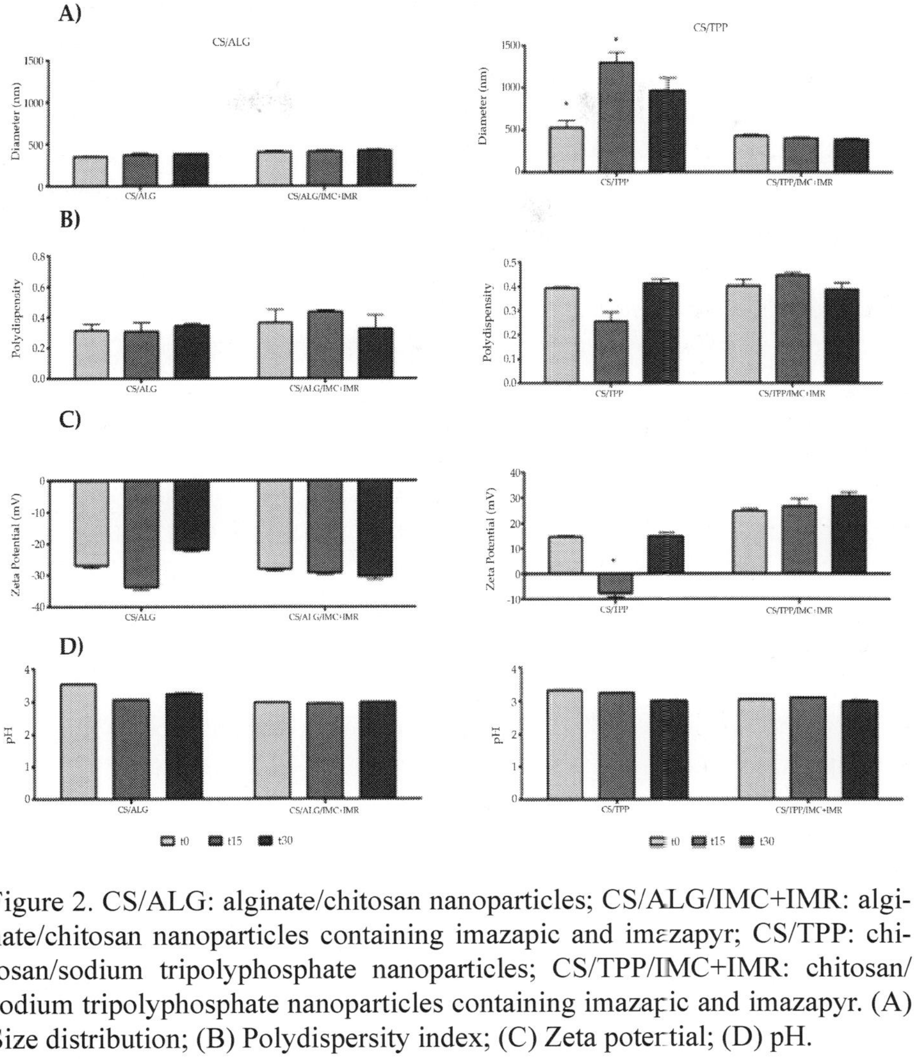

Figure 2. CS/ALG: alginate/chitosan nanoparticles; CS/ALG/IMC+IMR: alginate/chitosan nanoparticles containing imazapic and imazapyr; CS/TPP: chitosan/sodium tripolyphosphate nanoparticles; CS/TPP/IMC+IMR: chitosan/ sodium tripolyphosphate nanoparticles containing imazapic and imazapyr. (A) Size distribution; (B) Polydispersity index; (C) Zeta potential; (D) pH.

Both nanoparticle systems showed more than one particle population. The differences between the data obtained with the two techniques could be explained by the fact that in the NTA technique, the samples were diluted many times, which might have caused the rupture of aggregates, resulting in smaller average size distribution values, compared to the DLS data. Analysis of the CS/TPP and CS/ALG nanoparticles without the presence of the herbicides showed similar size distribution values (data not shown). The stabilities of the CS/TPP and CS/ALG nanoparticles, with or without the herbicides, were evaluated using measurements of average size distribution, polydispersity, zeta potential, and pH over a period of 30 days (Fig. 2).

In the case of the CS/ALG nanoparticles, no significant changes in these parameters were observed, indicating that the particles remained stable (in the presence and absence of the herbicides). An important finding is that the zeta potential remained at around − 30 mV, so that electrostatic repulsion was maintained between the particles, minimizing the likelihood of aggregation. In contrast, in the absence of the herbicides, the CS/TPP nanoparticles showed temporal changes in average size, polydispersity, and zeta potential that were indicative of particle aggregation, which was favored by the low zeta potential (+ 15 mV). However, the encapsulation of imazapyr and imazapic in the nanoparticles resulted in increased stability, with higher zeta potential (+ 26 mV). This could be explained by interactions between the herbicides and the polymeric chains of the CS/TPP nanoparticles. Comparison of the stabilities of the two types of nanoparticle containing herbicides showed that the polydispersity of the CS/ALG nanoparticles was lower than that of CS/TPP (Fig. 1).

The encapsulation efficiencies obtained for CS/ALG were 62.3 ±} 3.2% and 71.3 ±} 2.8% for IMC and IMR, respectively, while the corresponding values for CS/TPP were 58.6 ±} 4.5% and 69.6 ±} 5.5%, respectively. For both types of nanoparticles, greater encapsulation efficiency was obtained for IMR rather than IMC. In terms of chemical structure, the difference between these compounds is the presence of a –CH3 group in the aromatic ring of IMC, which slightly increases the hydrophobicity of this molecule and thereby decreasing the possibility of polar interactions with the polymers of the nanoparticles. Encapsulation efficiencies achieved using CS/ALG nanoparticles have been described in previous work. For instance, Sarmento *et al.* (2007) reported an insulin encapsulation efficiency of around 80%, which was dependent on the relative amounts of CS and ALG19. Similarly, Silva *et al.* (2010, 2011) reported encapsulation efficiencies for the herbicides clomazone (60 and 90%)20 and paraquat (75%)14.

The encapsulation efficiencies of IMC and IMR could be explained by the different strengths of intermolecular forces such as ionic interactions (pKa values of 3.9 for IMC and 3.6 for IMR) and hydrogen bonding (due to the presence of electronegative groups) with the components of the nanoparticles, especially the polymeric chains of chitosan. In the case of the CS/ALG nanoparticles, the presence of two polymers may have facilitated interaction with the herbicides, resulting in greater encapsulation efficiency. Nonetheless, an important point to consider is that the presence of a free herbicide fraction is required in order to provide immediate herbicidal action of the formulation.

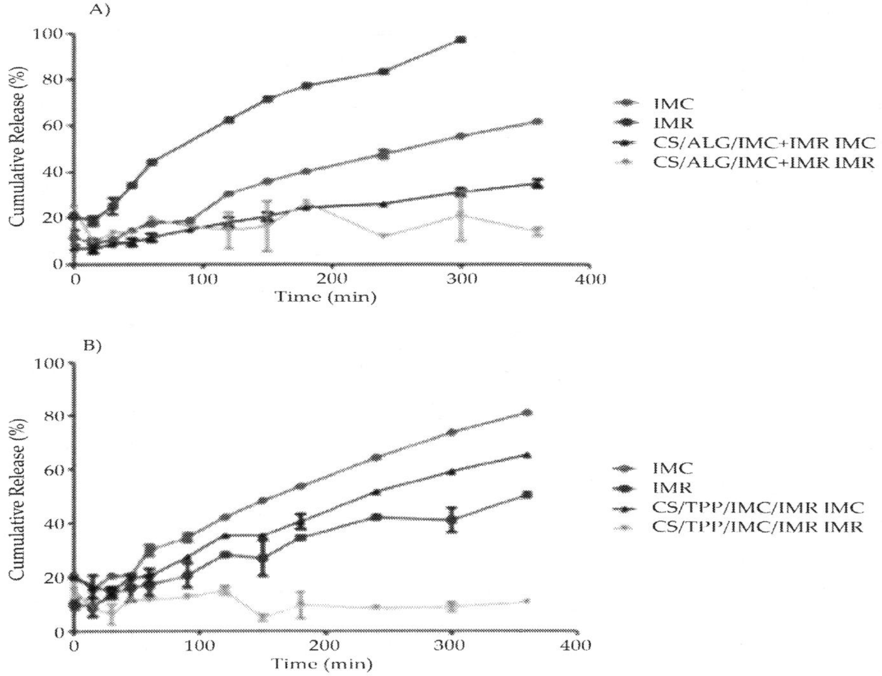

Figure 3. In vitro release kinetics profiles for imazapic and imazapyr associated with (A) CS/ALG nanoparticles, and (B) CS/TPP nanoparticles. Tests performed at 25 oC, in triplicate (n=3), using a system with two compartments (donor and acceptor) separated by a membrane with a molecular exclusion pore size of 1000 Da. The acceptor compartment contained solutions of 0.0022 mol/L calcium chloride (case A) or 0.0003 mol/L TPP (case B).

Assays of Release of the Herbicides from the Nanoparticles

The release profiles of the encapsulated combined herbicides from the CS/ALG nanoparticles are shown in Fig. 3. The release percentages of

55 and 97% were obtained for free IMC and IMR, respectively, after 300 min, while in the presence of the nanoparticles, the values decreased to 30% (IMC) and 20% (IMR) after 300 min. The results therefore showed that encapsulation of the combined herbicides resulted in slower release, compared to the free compounds When the herbicides were encapsulated separately in the CS/TPP nanoparticles, release percentages of 59 and 9% were obtained for IMC and IMR, respectively, after 300 min (Fig. 3). Combined encapsulation of the compounds in the CS/TPP nanoparticles resulted in a substantial decrease (of around 35%) in the release percentage of IMR, compared to separately encapsulated IMR, while the combined encapsulation did not affect the release of IMC.

Release mechanisms. The Higuchi, Korsmeyer-Peppas, and first order mathematical models were used to investigate the mechanism of release of IMC and IMR encapsulated together in the CS/ALG or CS/TPP nanoparticles. Figure 4 shows the results obtained from linearization of the release curves for IMC and IMR, encapsulated individually or in combination. Linear regression was used to calculate the values of the release constant and release exponent (Table 1). For the herbicides encapsulated together in the CS/ALG nanoparticles, the values obtained indicated that the release of IMC followed a diffusion mechanism according to Fick's law, while IMR was released by means of Case II transport, reflecting a different mode of interaction between the herbicide and the nanoparticles, as well as the lower encapsulation efficiency of IMR.

Silva *et al.* (2011) found significant differences between the release profiles of free and encapsulated paraquat14. The Korsmeyer-Peppas mathematical model was able to describe the behavior of the formulations, and release coefficient of 0.83 indicated that the release mechanism was by means of non-Fickian diffusion. The release parameter values obtained by fitting of the mathematical models to the data for the CS/TPP nanoparticles (Table 1) showed that the release of IMC was by diffusion following Fick's law, described by the Higuchi model.

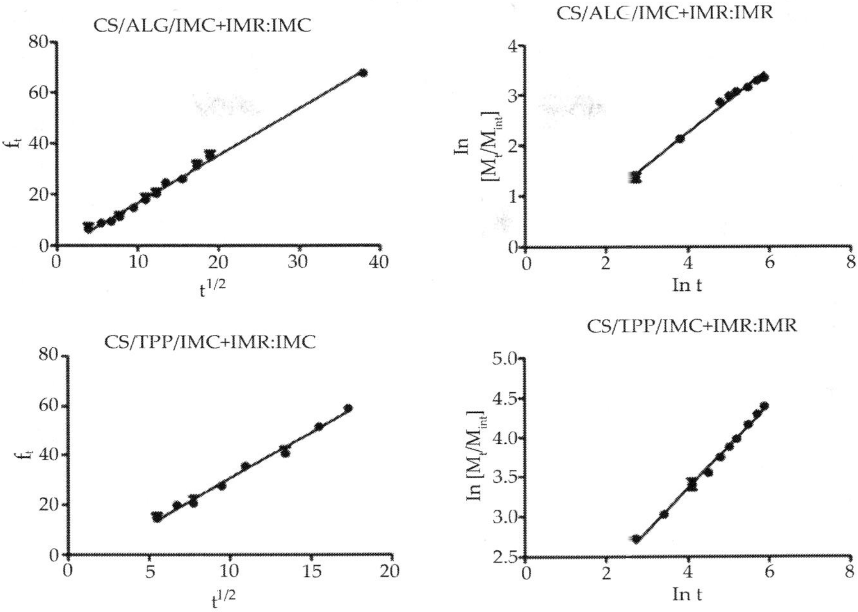

Figure 4: Linearization using the mathematical models that provided the best fits (Higuchi and Korsmeyer-Peppas) for the herbicides IMC and IMR encapsulated individually or in combination in the CS/ALG and CS/TPP nanoparticles.

Table 1: Parameter values obtained by fitting of mathematical models to the data for the CS/ALG and CS/TPP nanoparticles containing herbicides.

Parameter	Korsmeyer-Peppas	Higuchi	**First order**
CS/ALG/IMC+IMR:IMC			
Release constant (k)	1.689 min^{-1}	1.760 min^{-1}	2.13×10^{-3} min^{-1}
Release exponent (n)	0.5564	–	–
Correlation coefficient (r)	0.9892	0.9957	0.8920
CS/ALG/IMC+IMR:IMR			
Release constant (k)	1.578 min^{-1}	11.08 min^{-1}	2.78×10^{-3} min^{-1}
Release exponent (n)	1.283	–	–

Correlation coefficient (r)	0.9877	0.4819	0.9079
CS/TPP/IMC+IMR:IMC			
Parameter	Korsmeyer-Peppas	Higuchi	First order
Release constant (k)	7.800 min^{-1}	6.789 min^{-1}	1.82×10^{-3} min^{-1}
Release exponent (n)	1.671	–	–
Correlation coefficient (r)	0.1684	0.9897	0.9607
CS/TPP/IMC+IMR:IMR			
Parameter	Korsmeyer-Peppas	Higuchi	First order
Release constant (k)	1.881 min^{-1}	7.407 min^{-1}	1.95×10^{-4} min^{-1}
Release exponent (n)	1.230	–	–
Correlation coefficient (r)	0.9948	0.3559	0.1948

The release of IMR obeyed the Korsmeyer-Peppas mathematical model, with a release exponent of 1.23 being indicative of release by relaxation of the polymeric matrix and/or particle dissolution. Grillo *et al.* (2014a, 2014b) reported an alteration in the release profile of paraquat after encapsulation, and that the release profile of atrazine was modified when the particles were coated with chitosan8,17. Application of the Korsmeyer-Peppas mathematical model to the release kinetics data gave values for the release constant (k), the release exponent (n), and the correlation coefficient (r) of 1.99 min−1, 0.68, and 0.984, respectively. The value of the release exponent, which provides an indication of the type of release mechanism, was in the range $0.43 < n < 0.85$, indicating that the process was controlled by anomalous transport (a combination of diffusion and Case II transport).

Evaluation of Nanoparticle Toxicity

Allium cepa. The *Allium cepa* assay provides an indication of the ability of a substance to cause alterations in the genetic material. The results obtained here demonstrated that encapsulation of the herbicides was able to reduce the extent of damage, compared to the free compounds (Fig. 5), indicating that the encapsulation procedure diminished the genotoxicity (the ability to cause alterations in the DNA) of the formulations. The free herbicides (IMC+ IMR) showed a relative alteration index value that was around 100% higher than for the herbicides encapsulated in the CS/ALG or CS/TPP nanoparticles. The lowest alteration index value was found for encapsulation in the CS/ALG nanoparticles. In the absence of the herbicides, the CS/ALG nanoparticles showed the same relative alteration index value as the negative control, while an increase of approximately 50% was observed for CS/TPP.

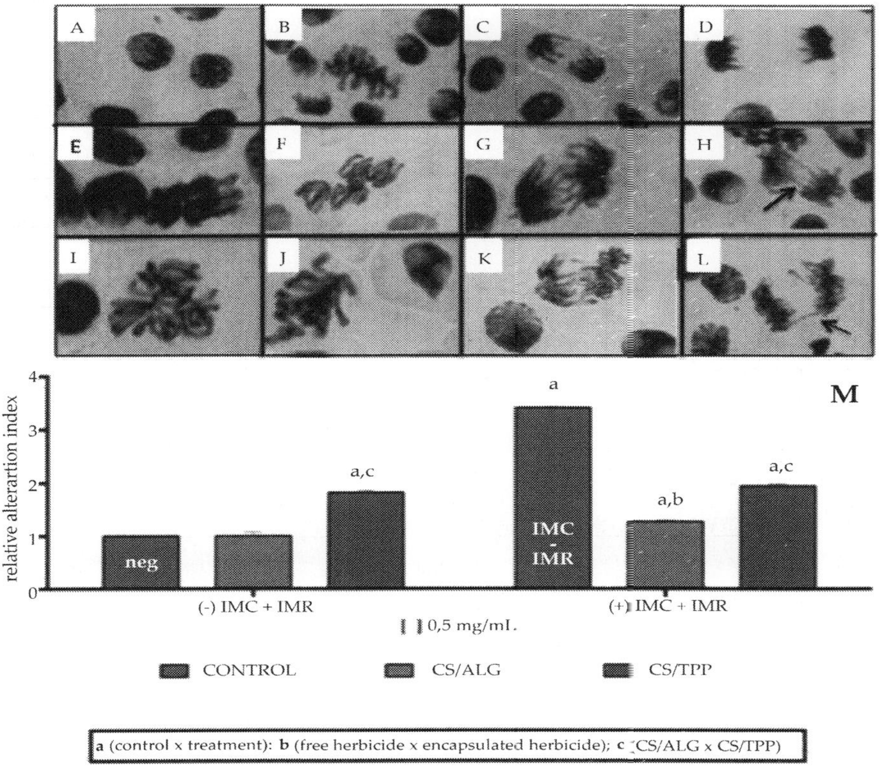

Figure 5. A to D cell division phases: (A) prophases; (B) metaphases; (C) anaphases; (D) telophase. In (E) to (L) show different alterations in division phases. (M) Relative alteration index values using the Allium cepa assay. The roots were exposed to the free herbicides, the CS/ALG and CS/TPP nanoparticles, and the herbicides encapsulated in the nanoparticles. Experiments performed

in triplicate at a concentration of 0.5 mg/mL. The letters a, b, and c indicate significant differences (ANOVA, $p < 0.05$).

For the encapsulated formulations, a smaller alteration index value was obtained using CS/ALG, which was therefore the best option in terms of reducing genotoxicity. The results therefore showed that encapsulation using CS/ALG provided slower release than the other formulations tested, and that the slower release resulted in less cellular damage over a period of 24 h.

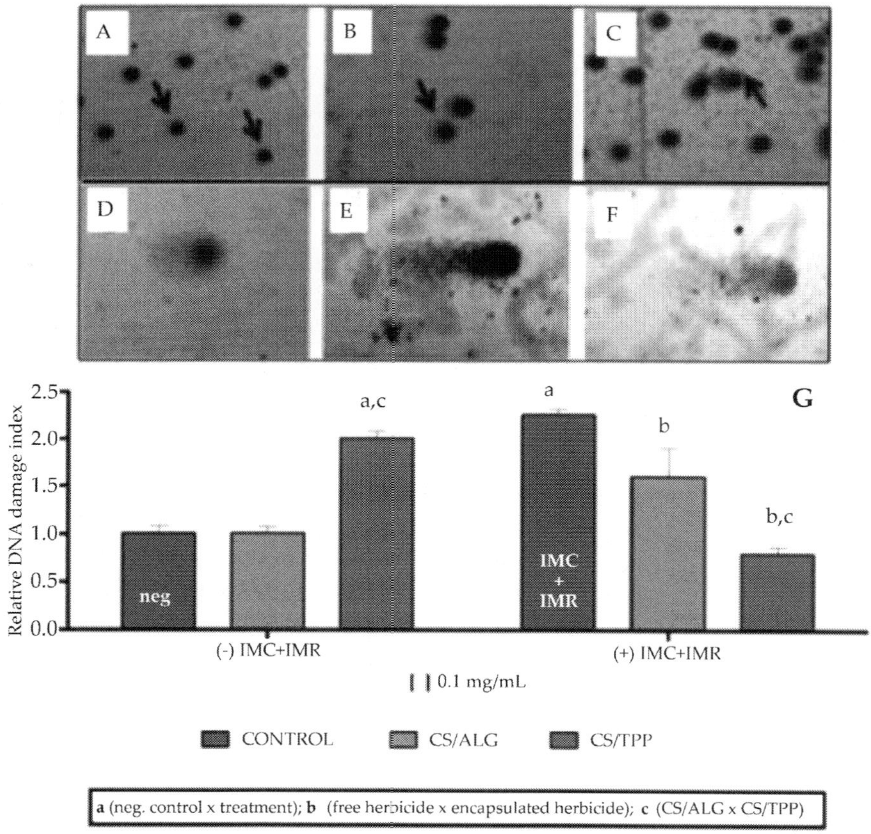

Figure 6: In (A) level alteration 0; in (B) level alteration 1; (C) level alteration 2; (D) level alteration 3; (E) level alteration 4 and (F) apoptosis. In (G) Relative damage index values obtained using comet tests with CHO cells exposed to the free herbicides, the CS/ALG and CS/TPP nanoparticles, and the herbicides encapsulated in the nanoparticles, at concentrations of 0.1 and 0.05 mg/mL.

In previous work, Grillo *et al.* (2014a) evaluated the genotoxic potential of free and encapsulated paraquat. The results showed that all the treatments increased DNA damage, compared to the negative control, although the greatest damage was caused by free paraquat, indicating that encapsulation of the herbicide acted to reduce damage to the DNA8. In work using atrazine, Grillo *et al.* (2010) showed that its encapsulation in PHBV reduced the toxic effects, compared to the free herbicide, because of the slower release15. *Comet assays.* The use of comet assays provides an indication of the capacity of a compound to rupture the genetic material, although it is a pre-mitotic analysis, which means that subsequent repair of the DNA might be possible during division. The comet analysis results showed that the use of the free herbicides resulted in higher damage index values, compared to the herbicides encapsulated in the CS/ALG and CS/TPP nanoparticles (Fig. 6).

The relative damage index value obtained for the CS/ALG nanoparticles without herbicides was similar to that for the control. However, in the case of the CS/TPP nanoparticles, the value obtained was higher than for both the CS/ALG nanoparticles and the negative control. The free herbicides showed the highest damage index value, which was 50% higher than that of the control. Encapsulation of the herbicides in the CS/TPP or CS/ALG nanoparticles reduced the extent of damage, with the best results obtained for CS/TPP+ herbicides, for which the index value was the same as that for the negative control. Nonetheless, no significant differences in genotoxicity were found between the negative control and the herbicides encapsulated using the two types of nanoparticle.

Similar findings were reported in previous *in vivo* studies using the same herbicides, with low toxicity to rats, dogs, and rabbits2. Grillo *et al.* (2012) compared DNA damage caused by treatments with triazine herbicides, free or encapsulated in poly-epsilon-caprolactone nanocapsules, and found that encapsulation resulted in reduced DNA damage16. Lima *et al.* (2012) studied the herbicide ametryn and found that encapsulation in microparticles of poly-hydroxybutyrate (PHB) and poly-hydroxybutyrate-co-hydroxyvalerate (PHBV) reduced DNA damage21.

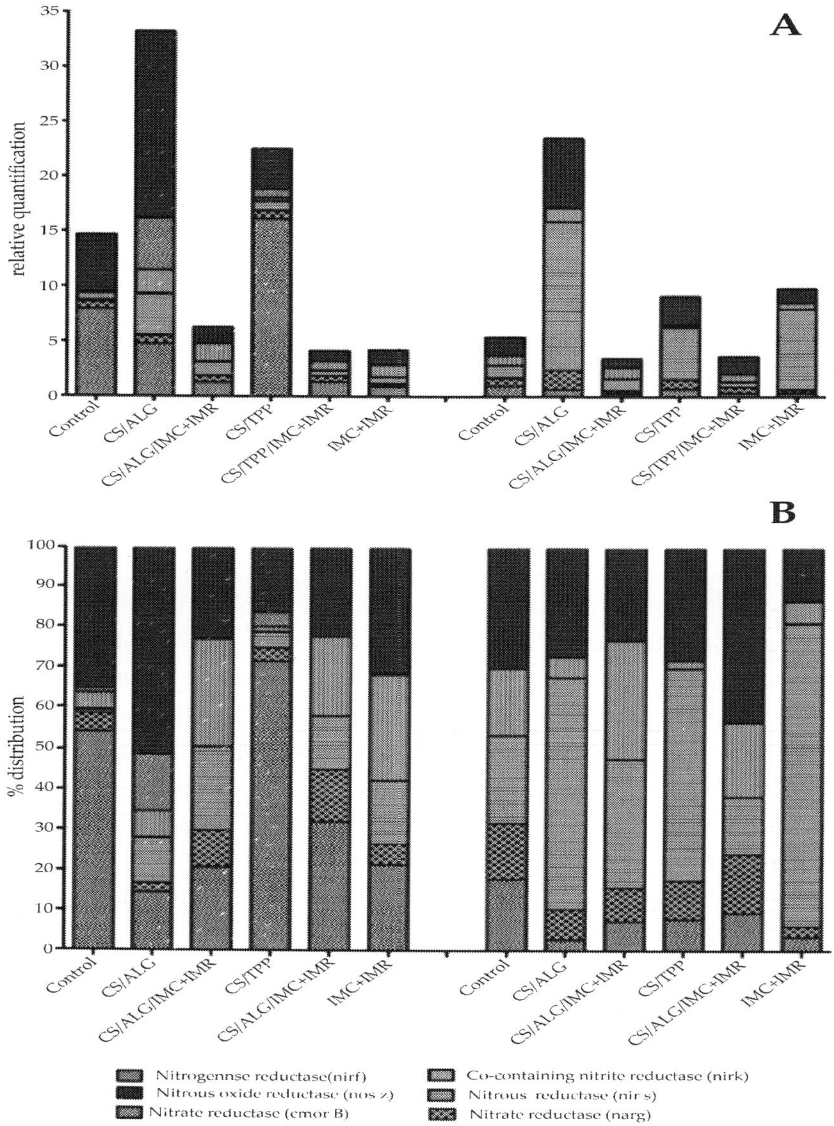

Figure 7. Quantitative soil analyses using the nirF, nosZ, cmorB, nirK, narG, and nirS genes: (A) 7 days after treatment; (B) 30 days after treatment. Proportions of genes present in the soil samples: (A) 7 days after treatment; (B) 30 days after treatment.

Analysis of the Effects of the Herbicides and Nanoparticles Containing the Herbicides on Soil Microbiota.

Quantification of the bacterial genes responsible for the production of enzymes active in the soil nitrogen cycle enabled assessment of the effects of the herbicides in their free and encapsulated forms on the numbers and proportions of bacteria responsible for different stages of the cycle. This information is important, because a properly functioning nitrogen cycle is essential for soil fertility.

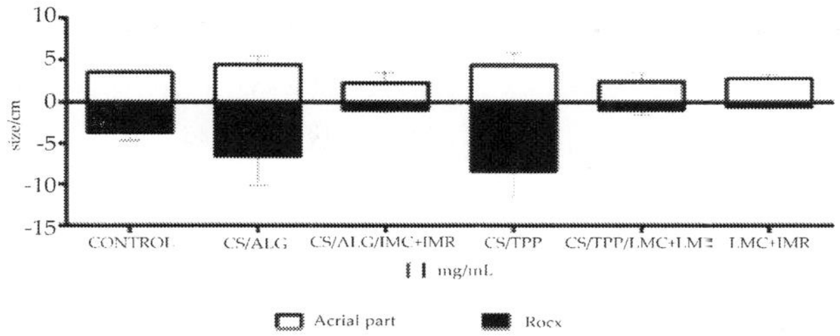

Figure 8: Effects of the CS/ALG and CS/TPP nanoparticles and the encapsulated herbicides (CS/ALG/IMC+IMR and CS/TPP/IMC+IMR), at concentrations equivalent to 400 g/ha, on the growth of Bidens pilosa.

The genes studied in the present work corresponded to nitrogen fixing, nitrifying, and denitrifying bacteria. Many studies have shown associations between environmental conditions, changes in the microbial community, and soil characteristics, considering aspects including contaminants, nutrients, water availability, and type of soil, amongst others22,23. Genes have been used to evaluate the cycles of nitrogen, carbon, phosphorus, and sulfur24–27.

Studies of the soil microbiota and their activities contribute to understanding the structure of the microbial community and the enzymatic activity involved in the cycles of elements in the soil. The present work focused on quantification of certain bacterial genes responsible for the synthesis of enzymes related to the nitrogen cycle. The evaluation was made relative to an initial control soil and a negative control soil that was monitored until the end of the experiment. Both of these soils remained untreated. The results revealed an increase in the numbers of bacteria in the soils treated with nanoparticles (CS/ALG and CS/TPP) (Fig. 7). This was mainly evident 7 days after application of the treatments, with an increase in the number of bacteria possessing the gene for synthesis of the nitrate reductase enzyme. Treatment of the soil with the free and encapsulated herbicides resulted in decreases in bacterial numbers, with the least alterations, compared to the negative control, shown by the soil exposed to the CS/ALG/IMC+ IMR formulation. After 30 days, higher numbers of bacteria (compared to the negative control) were observed for the soils treated with the nanoparticles alone (CS/ALG and CS/TPP), while the treatment with CS/ALG/IMC+ IMR showed the greatest similarity to the negative control.

Calculation of the percentage fractions of the genes related to the nitrogen cycle were used to determine the proportions of the different bacterial types present (Fig. 7B). The bacteria responsible for reduction of NO to N2O by the nitrate reductase enzyme constituted the largest fraction in the treatment with CS/ALG, while treatment with CS/TPP resulted in the greatest increase in the proportion of bacteria that produce the nitrogenase reductase enzyme responsible for the fixation of atmospheric nitrogen. This finding can be explained by the chemical composition of the nanoparticles, because the phosphate in TPP is likely to have a positive effect on bacterial growth, hence facilitating the process of nitrogen fixation28. The proportions of bacteria in the soil treated with the free herbicides were similar to those of the negative control, although the absolute numbers of bacteria were lower.

After 30 days, the treated soils showed distributions of bacteria similar to that of the control soil, with recovery shown by the soil treated with the herbicides (IMC+ IMR), in terms of the total number of bacteria. However, there was a general increase in the number of bacteria responsible for denitrification, especially the first stage (nitrite and nitrate reductases – *nirK*, *nirS*, and *narG* genes), as well as the second stage (nitrate reductase and nitric oxide reductase – *cnorB* and *nosZ* genes), although the numbers of bacteria involved in the second step were smaller for the herbicide-treated soils. When the concentration of denitrifying bacteria involved in Stage 1is smaller than the concentration for Stage 2, there is likely to be release of nitrogen to the environment. An abundance of bacteria presenting the copper-containing nitrite reductase gene (*nirK*), together with those presenting the nitrite reductase gene (*nirS*), results in increased degradation of NO2 − and greater NO formation. It should be noted that the nanoencapsulation systems altered the release profiles of the herbicides, with slower release during the first 7 days, so that the herbicidal activity outside the nanoparticles was lower, compared to the high initial level of activity provided by the free herbicides.Changes over time in the bacterial community responsible for the soil nitrogen cycle can be caused by physical factors such as heat, water availability, and pH29–34. Wang *et al.* (2014) used the qPCR technique and specific genes associated with the soil nitrogen cycle to identify alterations in the soil microbiota and ecological processes in different types of fertilized soils33.

The alterations observed up to 30 days indicated that imazapic and imazapyr induced changes in the soil microbiota in terms of both the quantity and types of the bacteria involved in the soil nitrogen cycle, with a decrease in the proportion of nitrogen-fixing organisms and an increase in denitrifying bacteria (especially those involved in the first step of denitrification). The soils treated with the encapsulated herbicides showed different microbiota profiles, compared to soils treated with the free herbicides. The bacterial profiles of the soils treated with the CS/ALG/IMC+ IMR and CS/TPP/IMC+ IMR formulations showed similarity to the negative control, especially in the case of CS/ALG/IMC+ IMR. This indicated that the carrier systems employed in the present work acted to reduce changes in the soil nitrogen cycle bacteria, compared to use of the free herbicides.

Evaluation of Herbicidal Activity.

The results described above demonstrated that encapsulation of the herbicides in the CS/TPP and CS/ALG nanoparticles led to a reduction in toxicity, indicating that the formulations could be attractive options for use in agriculture. The effectiveness of the encapsulated herbicides was therefore tested against a target weed species, *Bidens pilosa* (black-jack). In this assay, the parameters evaluated were the fresh masses of the roots and aerial parts of the plants, and the herbicides were applied at a dosage equal to that used in the field (400 g/ha). Treatment using both the free and encapsulated herbicides resulted in reduced growth, compared to the control (Fig. 8). The findings indicated that encapsulation should result in a reduction of toxicity of the herbicides towards nontarget organisms, while maintaining adequate herbicidal activity. Similarly, Grillo *et al.* (2014a) found that plants treated with CS/TPP nanoparticles containing paraquat showed lower growth than plants that received the free herbicide8.

CONCLUSIONS

The results of this work showed that CS/ALG and CS/TPP nanoparticles were able to encapsulate the herbicides imazapic and imazapyr with efficiencies exceeding 60%. Release kinetics experiments demonstrated that the encapsulated herbicides were released more slowly than the free forms. In the case of imazapic, the mechanism of release from both types of nanoparticle was by simple diffusion, while imazapyr was released from the CS/ALG nanoparticles by anomalous transport and from the CS/TPP nanoparticles by relaxation of the polymeric matrix. The nanoparticle size was smaller than 400 nm, the polydispersity index was around 0.3, and the zeta potentials of the formulations employing CS/ALG and CS/TPP were − 30 and + 26 mV, respectively. The formulations remained stable for 30 days.

Cytotoxicity assays indicated that the nanoparticle systems presented low toxicity. Genotoxicity assays using different cell types (CHO and *Allium cepa*) showed that the encapsulated herbicides caused less damage, compared to the free compounds. Tests using soil microbiota revealed that the free and encapsulated herbicides did not substantially affect the total numbers of bacteria present in the soil. However, molecular analysis using specific genes showed that there were changes in the amounts and types of bacteria associated with the soil nitrogen cycle, with the least interference shown by the herbicides encapsulated in the CS/ALG nanoparticles.

In tests of herbicidal activity, the encapsulated compounds were found to be more effective than the free forms, enabling the use of smaller dosages. In terms of reduced toxicity and herbicidal effectiveness, the best results were obtained using the CS/ALG nanoparticles. This formulation therefore offers an effective alternative for weed control in agricultural cultivations. Further tests are underway to evaluate the toxicity and herbicidal effectiveness of formulations containing lower concentrations of the herbicides.

The present study underlines that the encapsulation of two herbicides in one carrier system to improve the activity and reduces the impacts to humans and environment is a good strategy for the sustainable agricultural production. However, more research in this field is need in order to scale-up this technology for largescale applications and also for underpinning the mode of action of these nanopesticides in plants at molecular level.

METHODS

Preparation of the nanoparticles. *Alginate/chitosan nanoparticles*. The alginate/chitosan (CS/ALG) nanoparticles were manufactured using the ionotropic gelification method described by De and Robinson (2003)35. A 10 mL solution of sodium alginate was prepared, to which was added the herbicides imazapic (IMC) and imazapyr (IMR). A solution of calcium chloride (0.24 mg/mL) was then added dropwise to the first solution.

The resulting calcium alginate pre-gel was maintained under agitation for 30 min. Subsequently, a solution of chitosan (0.24 mg/mL) in 1% acetic acid, previously prepared and kept under agitation for 12 h to solubilize the polymer was added and the mixture was kept under agitation overnight to enable the formation of nanospheres. The final concentrations of imazapyr and imazapic in the formulation were 1 g/mL.

Chitosan/tripolyphosphate nanoparticles. For the production of chitosan/tripolyphosphate (CS/TPP) nanoparticles, a 0.2% solution of chitosan in 0.6% acetic acid was prepared and kept under agitation for 12 h. The solution was then diluted to 0.1% chitosan in 0.3% acetic acid, and the herbicides were added. The final solution volume was 10 mL. In the next step, 6 mL of a 0.1% solution of TPP (at pH 4.5 and 8 oC) were added dropwise to the chitosan solution and the mixture was agitated for 12 h. The final concentrations of imazapir and imazapic in the formulation were 1 mg/mL.

Physicochemical Characterization

Characterization of the CS/TPP and CS/ALG nanoparticles was performed by means of size distribution measurements using dynamic light scattering (DLS) and nanoparticle tracking analysis (NTA). The DLS analyses employed a ZS90 instrument (Malvern Instruments, UK). The NTA measurements used 1 mL of nanoparticle suspension, with each measurement consisting of five repetitions with 90 s videos and an average of 100 particles per 1 s frame. The result was a graph of concentration according to nanoparticle size. All the DLS and NTA analyses were made in triplicate. The polydispersity and zeta potential of the nanoparticles were determined by microelectrophoresis, using a Zeta Plus analyzer (Malvern Instruments).

The stability of the suspensions, with and without the herbicides, was evaluated by measuring the average particle diameter, polydispersity, and zeta potential during a period of 30 days, with the samples stored in amber flasks at ambient temperature (25 °C).

Efficiency of Encapsulation of the Herbicides in the Nano-particles

The total amount (100%) of herbicide present in the polymeric nanoparticle suspension was determined by dilution of the suspensions in methanol. This solution was filtered through a 0.22 μ m membrane (Millipore), and quantification was performed using high performance liquid chromatography (HPLC). The amounts of herbicides encapsulated in the nanoparticles were determined using the ultrafiltration/centrifugation method, with the nanoparticle suspensions centrifuged in 30 kDa regenerated cellulose ultrafiltration devices (Microcon, Millipore). In this procedure, only the herbicides were able to cross the membrane, and the ultrafiltrate was analyzed by HPLC. The amounts of herbicide associated with the nanoparticles could then be obtained from the difference between the total amounts (100%) and the amounts not associated with the nanoparticles36–38.

Release Kinetics Assays

Assays to measure the release profiles of the herbicides were performed using a system with two compartments (donor and acceptor), maintained under gentle agitation15. A cellulose membrane (1 kDa, Spectrapore) was used to separate the sample (2 mL) in the donor compartment from the acceptor compartment. The pores of the membrane only allowed passage of the free herbicides, with the herbicides associated with the nanoparticles being retained. Samples were collected from the acceptor compartment for HPLC analysis after different time intervals. All the measurements were performed in triplicate. The mechanisms of release of the herbicides were evaluated using the first order, Higuchi, and Korsmeyer-Peppas mathematical models39–41.

Cytotoxicity and Genotoxicity Assays

Allium cepa assays. In these tests, the roots of germinated seeds were placed for 24 h in the different media (containing nanoparticles, nanoparticles with herbicides, and herbicides, at concentrations of 0.5 mg/mL). After this period, the roots were fixed in methanol:acetic acid (3:1) for 24 h, followed by treatment with 1 mol/L HCl for 9 min at 60 °C. The samples were dyed with Schiff reagent for 2 h and then spread onto slides, with the addition of one drop of 2% acetic-carmine. The slides were examined using an optical microscope at a magnification of x40. The results were used to calculate the mitotic index (MI) and the damage index (DI).

Comet tests. Comet tests were performed using Chinese hamster ovary (CHO) cell cultures. The CHO cell line was maintained in bottles containing McCoy culture medium (Gibco) supplemented with 10% fetal bovine serum and 1% antibiotic, kept in a heating cabinet at 37 °C under a humid atmosphere with 5% CO2, until reaching a satisfactory number of cells for use in the assays. The cells were then subcultured and subsequently transferred to 6-well plates, followed by treatment during 1 h with the CS/ALG and CS/TPP nanoparticles, with and without the herbicides. Negative controls were included, without any treatment. After a period of 1 h, the cells (treated and controls) were separately mixed with low melting agarose (0.8%) at 37 °C, and the mixture was spread onto a slide that had been pre-coated with normal agarose (1.5%). Cover slips were placed over the samples and the submerged in lysis solution for 1 h, followed by 5 min in neutralization solution (Tris). After neutralization, the slides were transferred to a horizontal electrophoresis cuvette containing buffer and allowed to rest for 20 min before starting the run, which was performed for 20 min. The slides were again neutralized for 10 min, then washed in water for 5 min and dried overnight at ambient temperature. The samples were subsequently rehydrated, dried again in a drying cabinet at 37 °C, transferred to fixing solution (containing trichloroacetic acid, zinc sulfate, and glycerol) for 10 min, and dried at 37 °C. In the next step, the slides were hydrated for 5 min in distilled water, stained with silver nitrate for 30 min, placed in stop solution (acetic acid) for 5 min, and washed three times with distilled water. The final drying was performed at ambient temperature, and the comet analysis employed an optical microscope at x40 magnification, with around 100 randomly selected cells examined for each sample. The damage indices were calculated according to the sizes of the tails. The tails were scored in five levels (0, 1, 2, 3 and 4), where zero and four represent respectively the lower and higher comet tails size (Eq1). A specialist analyzed the slides using a single-blind-review in order to minimize variability.

DI = (_***cel*** 0 ***x*** 0) + (_***cel*** 1 ***x*** 1) + (_***cel*** 2 ***x*** 2) (_***cel*** 3 ***x*** 3) + (_***cel*** 4 ***x*** 4)/ ***total cell*** (1)

Table 2: Analysis of soil composition using X-ray fluorescence.

Element	Concentration (mg/cm^2)	Element	Concentration (mg/cm^2)
Aluminium	0.832	Titanium	0.755
Calcium	1.675	Chromium	0.004
Chloro	0.174	Manganese	0.029
Iron	5.654	Zinc	0.007
Potassium	0.454	Strontium	0.008
Magnesium	0.252	Ytrium	0.002
Phosphurus	0.004	Zirconium	0.087
Silicon	15.007	Arsenium	0.000
Sulfur	0.049	Bromine	0.000

Analysis of the Effects of the Herbicides and the Herbicide-loaded Nanoparticles on Soil Microbiota.

The soil used in this experiment was obtained from a local agricultural supplier. The chemical composition of the soil was characterized using X-ray fluorescence, considering the elements aluminum, calcium, chlorine, iron, potassium, magnesium, phosphorus, silicon, sulfur, titanium, chromium, manganese, zinc, strontium, yttrium, zirconium, arsenic, and bromine (Table 2). A 30 g quantity of soil was weighed and sieved for each treatment (negative control, CS/ALG, CS/ALG/IMC+ IMR, CS/TPP, CS/TPP/IMC+ IMR, and IMC+ IMR). A single exposure to the test media was performed, after which the soils were watered twice weekly in order to maintain constant humidity. The soil was placed in containers, with surface area of 0.025 m2, and the quantities of herbicides used were equivalent to the application rates employed in the field, with a proportion of 75% imazapic to 25% imazapyr (using concentrations of 75 g/L of imazapic and 25 g/L of imazapyr). Extraction and analysis of the soil microbiota DNA was conducted 7 and 13 days after application of the treatments.

Molecular Analysis of the Soil Microbiota

Extraction of DNA from the soil. DNA extraction for the polymerase chain reactions employed a PowerSoil DNA Extraction Kit. The genetic material was quantified by fluorescence, using a Qubit 3.0 Fluorometer and a Qubit dsDNA BR Assay Kit (Invitrogen). After quantification, the extractions were diluted to a final total DNA concentration of 1000 ng/mL. *qPCR.* Quantification of the 16S rRNA, *nifH, nirk, nirS, narG (Escherichia coli), norB, nosZ, and narG* (*Pseudomonas aeruginosa*) bacterial genes in the soil samples treated with the nanoparticles (with and without herbicides) and the free herbicides was achieved using real-time polymerase chain reactions (qPCR). The primers employed are described in Table 3. The reactions were performed in a final volume of 25 μ L containing: 12.5 μ L of Platinum SYBR Green qPCR SuperMix-UDG with ROX (Invitrogen), 1 μ L of each primer (sense and anti-sense), 1 μ L of template (DNA previously extracted from the soil using a PowerSoil DNA Isolation Kit (MO BIO Laboratories, Inc.), and autoclaved ultrapure water to make up the volume to 25 μ L. The amplification followed the procedure described by Jung *et al.* (2011), with an initial denaturation at 95 °C for 3 min, followed by 40 cycles of 95 °C for 45 s, 60 °C for 45 s, and 72 °C for 45 s. The fluorescence was measured at the end of each incubation at 60 °C42.

A calibration curve was constructed using serial dilutions of the DNA (1:1, 1:10, 1:100, and 1:1000 v/v) and the slope was used to calculate the average amplification efficiency. The sample quantification results were expressed as Δ Δ CT, comparing the control and the treated samples.

Table 3: Primers used in the qPCR reaction for quantification of specific genes.

Gene	Primer	Sequence	pb	Reference
Bacterial 16S rRNA gene	341F 534R	5′ CCTACGGGAGGCAGCAG 3′ 5′ ATTACCGCGGCTGCTGGCA 3′	193	Watanabe et al., 2001[43]
Nitrogenase reductase; nifH	nifHF nifHRb	5′AAAGGYGGWATCGGYAARTCCACCAC 3′ 5′ TGSGCYTTGTCYTCRCGGATBGGCAT 3′	400	Rösch and Bothe, 2005[44]
Cu-containing nitrite reductase; nirK	nirK 1F nirK 5R	5′GGMATGGTKCCSTGGCA 3′ 5′ GCCTCGATCAGRTTRTGGTT 3′	514	Braker *et al.*, 1998[45]
Nitrite reductase; nirS	nirS cd3AF nirS R3cd	5′GTSAACGTSAAGGARACSGG 3′ 5′ GASTTCGGRTGSGTCTTGA 3′	425	Throback et *al.*, 2004[46]
Nitrate reductase; norB	cnorB2F cnorB6R	5′GACAAGNNNTACTGGTGGT 3′ 5′ GAANCCCCANACNCCNGC 3′	389	Braker and Tiedje, 2003[47]
Nitrous oxide reductase; nosZ	nosZ-F nosZ-R	5′CGYTGTTCMTCGACAGCCAG 3′ 5′ CGSACCTTSTTGCCSTYGCG 3′	453	Kloos *et al.*, 2001[48]
Nitrate reductase; narG	narG-f narG-r	5′TCGCCSATYCCGGCSATGTC 3′ 5′ GAGTTGTACCAGTCRGCSGAYTCSG 3′	173	Bru *et al.*, 2007[49]

Evaluation of Herbicidal Activity

The activities of imazapic and imazapyr (free or encapsulated), were evaluated using *Bidens pilosa* (black-jack) in 10 cm diameter pots. The soil used was a commercial product composed of pine bark, peat, vermiculite, class A organic agricultural waste, wood sawdust, manure, bone meal, magnesium thermophosphate, and castor oil cake. The experiment was performed in triplicate, using the free and encapsulated herbicides (IMC+ IMR), the two types of nanoparticle (CS/ALG and CS/TPP) without herbicides, and untreated controls. Ten *Bidens pilosa* seeds were planted in each pot, totaling 30 seeds per triplicate. The quantities of herbicide used were equivalent to the amounts used in the field (400 g/ha), with the treatments applied pre-emergence. The pots were filled with around 200 g of substrate, the seeds were planted, and the treatments were applied (IMC, IMR, CS/ALG, CS/ALG/IMC+ IMR, CS/TPP, and CS/TPP/IMC+ IMR). The pots were kept in a greenhouse and were watered every day (morning and afternoon) during 15 days. After this period, the plants were removed and the following parameters were analyzed: germination efficiency, height, root length, and green mass.

REFERENCES

1. Peres, F., Moreira, J. C. & Dubois, G. S. *Agrotóxicos, saúde e ambiente: uma introdução ao tema*. In: Peres, F. & Moreira, J. C. (Ed), 21–41 (Fiocruz, 2003).
2. Krieger, R. I. & Krieger, W. C. in Handbook of *Pesticide Toxicology. 2nd edn*, (Ed. Krieger, R. I.) Ch. 74, (Elsevier, 2001).
3. Dayan, F. E. & Zaccaro, M. L. M. Chlorophyll fluorescence as a marker for herbicide mechanisms of action. *Pest. Biochem. Physiol.* **102,** 189–197 (2012).
4. Ulbrich, A. V., Souza, R. P. & Shaner, D. Persistence and Carryover Effect of Imazapic and Imazapyr in Brazilian Cropping Systems. *Weed Technology*, **19,** 986–991 (2005).
5. Ghormade, V., Deshpande, M. V. & Paknikar, K. M. Perspectives for nano-biotechnology enabled protection and nutrition of plants. *Biotechnol Adv.* **29,** 792–803 (2011).
6. Khot, L. R., Sankaran, S. & Maja, J. M. Applications of nanomaterials in agricultural production and crop protection: A review. *Crop Protec*. **35,** 64–70, (2012).
7. Kah, M., Beulke, S., Tiede, K. & Hofmann, T. Nanopesticides: state of knowledge, environmental fate, and exposure modeling. *Crit. Rev. Env. Sci. Tec*. **43,** 1823–1867 (2013).
8. Grillo, R. *et al.* Chitosan/tripolyphosphate nanoparticles loaded with paraquat herbicide: An environmentally safer alternative for weed control. *J. Haz. Mat*. **278,** 163–171 (2014a).
9. Pereira, A. E. S., Grillo, R., Mello, N. F. S., Rosa, A. H. & Fraceto, L. F. Application of poly(epsilon-caprolactone) nanoparticles containing atrazine herbicide as an alternative technique to control weeds and reduce damage to the environment. *J. Haz. Mat*. **268,** 207–215 (2014).
10. Oliveira, J. L. *et al.* Solid Lipid Nanoparticles Co-loaded with Simazine and Atrazine: Preparation, Characterization, and Evaluation of Herbicidal Activity. *J. Agric. Food Chem*. **63,** 422–432 (2015).
11. Durán, N., Mattoso, L. H. C. & Morais, P. C. Nanotecnologia: introdução, preparação e caracterização de nanomateriais e exemplos

de aplicação. (Artliber, 2006).

12. Nguyen, H. M., Hwang, I. C., Park, J. W. & Park, H. J. Enhanced payload and photo-protection for pesticides using nanostructured lipid carriers with corn oil as liquid lipid. *J Microencapsul.* **29,** 596–604 (2012a).
13. Nguyen, H. M., Hwang, I. C., Park, J. W. & Park, H. J. Photoprotection for deltamethrin using chitosan-coated beeswax solid lipid nanoparticles. *Pest Manag Sci.* **68,** 1062–1068 (2012b).
14. Silva, M. S. *et al.* Paraquat-loaded alginate/chitosan nanoparticles: preparation, characterization and soil sorption studies. *J. Haz. Mat.* **190,** 366–374 (2011).
15. Grillo, R. *et al.* Characterization of Atrazine-Loaded Biodegradable Poly(Hydroxybutyrate-Co-Hydroxyvalerate) Microspheres. *J Polym Environ.* **18,** 26–32 (2010).
16. Grillo, R. *et al.* Poly(e-caprolactone)nanocapsules as carrier systems for herbicides: Physico-chemical characterization and genotoxicity evaluation. *J. Haz. Mat.* **231–232,** 1– 9 (2012).
17. Grillo, R., Rosa, A. H. & Fraceto, L. F. Poly(ε -caprolactone) nanocapsules carrying the herbicide atrazine: effect of chitosan-coating agent on physico-chemical stability and herbicide release profile. *Int. J. Environ. Sci. Technol.* **11,** 1691–1700 (2014b).
18. Kashyap, P. L., Xiang, X. & Heiden, P. Chitosan nanoparticle based delivery systems for sustainable agriculture. *Int. J. Biol. Macromol.* **77,** 36–51 (2015).
19. Sarmento, B., Ribeiro, A.J., Veiga, F., Ferreira, D.C. & Neufeld, R.J. Insulin-Loaded Nanoparticles are Prepared by Alginate Ionotropic Pre-Gelation Followed by Chitosan Polyelectrolyte Complexation. *J. Nanosci. Nanotechnol.* **7,** 2833–2841 (2007).
20. Silva, M. S. *et al.* Nanopartículas de alginato como sistema de liberação para o herbicida clomazone. *Quim Nova*, **33,** 1868–1873 (2010).
21. Lima, R., Feitosa, L. O., Grillo, R., Pereira, A. E. S. & Fraceto, L. F. Evaluation of the Genotoxicity of Polymeric Microparticles Containing Ametryn Herbicide. *J. Environ. Sci. Eng.* **4,** 553–565 (2012a).

22. Buckley, D. H. & Schmidt, T. M. Diversity and dynamics of microbial communities in soils from agro-ecosystems. *Environ. Microbiol.* **5,** 441–452 (2003).

23. Stark, C. H., Condron, L. M. & O'callaghan, M. Differences in soil enzyme activities, microbial community structure and short-term nitrogen mineralization resulting from farm management history and organic matter amendments. *Soil Biol. Biochem.* **40,** 1352–1363 (2008).

24. Liang, Y., Zhao, H., Zhang, X., Zhou, H. & Li, G. Contrasting microbial functional genes in two distinct saline-alkali and slightly acidic oil-contaminated sites. *Sci Total Environ.* **487,** 272–278 (2014).

25. Xing, Y., Si, Y.X., Hong, C. & Li, Y. Multiple factors affect diversity and abundance of ammonia-oxidizing microorganisms in iron mine soil. *Arch Environ Contam Toxicol.* **69,** 20–31 (2015).

26. Su, J.Q. *et al.* Long-term balanced fertilization increases the soil microbial functional diversity in a phosphorus-limited paddy soil. *Mol. Ecol.* **24,** 136–150 (2015).

27. Brankatschk, R., Fischer, T., Veste, M. & Zeyer, J. Succession of N cycling processes in biological soil crusts on a Central European inland dune. *FEMS Microbiol Ecol.* **83,** 149–160 (2013).

28. Barroti, G. & Nahas, E. População microbiana total e solubilizadora de fosfato em solo submetido a diferentes sistemas de cultivo. *Pesq. Agropec. Bras.* **35,** 2043–2050 (2000).

29. Bell, C. W., Acosta-Martinez, V. & Mcintyre, N. E. Linking microbial community structure and function to seasonal differences in soil moisture and temperature in a Chihuahuan desert grassland. *Microb. Ecol.* **58,** 827–842 (2009).

30. Rasche, F. *et al.* Seasonality and resource availability control bacterial and archaeal communities in soil of a temperate beech forest. I*SME J.* **5,** 389–402 (2011).

31. Hayden, H. L., Mele, P. M. & Bougoure, D. S. Changes in the microbial community structure of bacteria, archaea and fungi in response to elevated CO2 and warming in an Australian native grassland soil. *Environ. Microbiol.* **14,** 3081–3096 (2012).

32. Lauber, C. L., Ramirez, K. S. & Aanderud, Z. Temporal variability

in soil microbial communities across land-use types. *ISME J.* **7,** 1641–1650 (2013).

33. Wang, H. *et al.* Temporal changes in soil bacterial and archaeal communities with different fertilizers in tea orchards. *J Zhejiang Univ-Sci B.* **15,** 953–965 (2014).
34. Di, H. J., Cameron, K. C. & Podolyan, A. Effect of soil moisture status and a nitrification inhibitor, dicyandiamide, on ammonia oxidizer and denitrifier growth and nitrous oxide emissions in a grassland soil. *Soil Biol. Biochem.* **73,** 59–68 (2014).
35. De, S. & Robinson, D. Polymer relationships during preparation of chitosan: alginate and poly-L-lysine–alginate nanospheres. *J. Control. Release*, **89,** 101–112 (2003).
36. Gamisans, F. *et al.* Flurbiprofen-loaded nanospheres: analysis of the matrix structure by thermal methods. *Int. J. Pharm.* **179,** 37–48 (1999).
37. Schaffazick, S. R., Guterres, S. S., Freitas, L. L. & Pohlmann, A. R. Caracterização e estabilidade físico-química de sistemas poliméricos nanoparticulados para administração de fármacos. *Quim. Nova*, **26,** 726–737 (2003).
38. Kilic, A. C. *et al.* Preparation and characterization of PLGA nanospheres for the targeted delivery of NR2B-specific antisense oligonucleotides to the NMDA receptors in the brain. *J. Microencapsul.* **22,** 633–641 (2005).
39. Higuchi, T. Rate of release of medicaments from ointment bases containing drugs in suspension. *J. Pharm. Sci.* **50,** 874–875 (1961).
40. Korsmeyer, R.W. *et al.* Mechanisms of solute release from poros hydrophilic polymers. *Int. J. Pharm.* **15,** 25–35 (1983).
41. Costa, P. & Lobo, J. M. S. Modeling and comparison of dissolutions profiles. *Eur. J. Pharm. Sci.* **13,** 123–133 (2001).
42. Jung, J. *et al.* Change in gene abundance in the nitrogen biogeochemical cycle with temperature and nitrogen addition in Antarctic soils. *Res. Microbiol.* **162,** 1018–1026 (2011).
43. Watanabe, K., Kodama, Y. & Harayama, S. Design and evaluation of PCR primers to amplify bacterial 16S ribosomal DNA fragments used for community fingerprinting. *J. Microbiol. Methods.* **44,** 253–

262 (2001).

44. Rösch, C. & Bothe, H. Improved assessment of denitrifying, N2-fixing, and total community bacteria by terminal restriction fragment length polymorphism analysis using multiple restriction enzymes. *Appl Environ Microbiol.* **71,** 226–235 (2005).
45. Braker, G., Fesefeldt, A. & Witzel, K.P. Development of PCR primer systems for amplification of nitrite reductase genes (nirK and nirS) to detect denitrifying bacteria in environmental. *Appl Environ Microbiol.* **64,** 3769–3775 (1998).
46. Throback, I.N., Enwall, K., Jarvis, A. & Hallin, S. Reassessing PCR primers targeting nirS, nirK and nosZ genes for community surveys of denitrifying bacteria with DGGE. *FEMS Microbiol Ecol.* **49,** 401–417 (2004).
47. Braker, G. & Tiedje, J.M. Nitric Oxide Reductase (***norB***) genes from pure cultures and environmental samples. *Appl Environ Microbiol.* **69,** 3476–3483 (2003).
48. Kloos, K., Mergel, A., Rosch, C. & Bothe, H. Denitrification within the genus Azospirillum and other associative bacteria. *Aust J Plant Physiol.* **28,** 991–998 (2001).
49. Bru, D., Sarr, A. & Philippot, L. Relative abundance of proteobacterial membrane-bound and periplasmic nitrate reductase in selected environments. *Appl Environ Microbiol.* **73,** 5971–5974 (2007).

SECTION 2

12

CHAPTER

SYNTHESIS AND CHARACTERIZATION OF NEW BIODEGRADABLE CHITOSAN/POLY-VINYL ALCOHOL/CELLULOSE NANOCOMPOSITE

Alireza Samzadeh-Kermani, Negin Esfandiary

Department of Chemistry, Faculty of Science, University of Zabol, Zabol, Iran

ABSTRACT

In this report, a new nanocomposite based on chitosan/polyvinyl alcohol/nanocrystalline cellulose (Cts/PVA/NCC) was synthesized. The morphology and particle size of NCC and nanocomposites were studied by scanning electron microscopy (SEM), X-ray diffraction (XRD) analysis and Fourier transform infrared (FT-IR) spectroscopy. According to XRD results, the size of NCC was found to be at the range of 15 - 17 nm. SEM images showed the rod-like shape of NCC whiskers. Finally, biodegradation and swelling studies were performed on Cts/PVA/NCC nanocomposites.

Keywords: Nanocomposite, Nanocrystalline Cellulose, Chitosan, Polyvinyl Alcohol, Biodegradation

INTRODUCTION

Nanostructures are produced from nanofillers which are dispersed in a polymer matrix. Incorporation of small amounts of nanometer-sized fillers can yield composites with enhanced properties earnestly required for many industrial and technological applications [1] . Naturally biopolymers, like polysaccharides can be made into either coatings or stand-alone films [2] . Many studies on biopolymer films for food packaging applications have been reported [3] . The production of NCC has gained increasing attention in recent years. Cellulose, one of the world's most abundant, natural and renewable biopolymer resources, is widely present in various forms of biomass, such as plants. In cellulosic plant fibers, cellulose is present in an amorphous form, but is associated with crystalline phases through both intermolecular and intramolecular hydrogen bondings in which cellulose does not melt before thermal degradation [4] . Cellulose is organized in fibrils, which are aligned parallel to each other, surrounded by a matrix of lignin and hemicellulose. The properties of cellulose including mechanical state, low density and biodegradability [5] depend on the type of cellulose which is present. Cellulose macromolecule consists of β-1, 4-D-linked glucose chains with molecular formula $(C_6H_{10}O_5)_n$ through an acetal oxygen which covalently bonds C-1 of one glucose ring to C-4 of the other [6] . Owing to the abundance of hydroxyl groups existent on the surface of cellulose nanocrystals, reactive NCC can be modified with different chemical groups to accomplish the expected surface modification like esterification and silylation or polymer grafting, which could successfully functionalize NCC and facilitate its dispersion into different polymer matrices [7] . Therefore, NCC is considered as one of the ideal nano-reinforcing agents for polymer matrices (including water soluble and water-insoluble polymer systems) and has been used into many polymer matrices to produce reinforced nanocomposites [8] . In addition, low density, low energy consumption, inherent renewability, biodegradability and biocompatibility are also good advantages of environmentally-friendly NCC [9] . Because of good dispersion of NCC in water [10] , fabrication and application of hydrogels including NCC without modification have many advantages versus other nanofillers such as polymer and metal nanoparticles [11] . The improved interface between nanofillers and polymer matrix is beneficial to the properties of polymer-based nanocomposites [12] . In nanocomposites, hydrogen bonding between NCC and polymer matrix plays an important role in determining polymer-NCC interaction. Strong hydrogen bonding between PVA and NCC was confirmed by observing the band gap between 3550 and 3200 cm^{-1} in Fourier transform infrared spectra (FTIR). In addition to hydrogen bonding, covalent bonding also provides an enhancement in polymer-nanofiller interface to achieve optimal nanobiocomposite properties [13] . NCC was made in spherical state [14] [15] , or rod-like highly crystalline nanocrystals [6] [16] -[18] , which was obtained by acid hydrolysis of cellulosic fibers and microfibrillated cellulose (MFC) resulting from disintegration of cellulose fibers under high shearing and impact forces [19] . NCC, which is obtained via acid hydrolysis of cellulose, typically has a wide range in size. With increasing consumption consciousness about petroleum-based polymer materials, the development of fully biodegradable, eco- friendly and bio-based nanocomposites has gained more and more attention both in industrial and academic fields [20] . The so-called "green" composites are derived from different natural resources such as plants or animals [21] . PVA is a material with technological potential which has a wide range of applications due to its unique physical and chemical properties [22] . It is a nontoxic, highly crystalline matter and has good film- forming and high hydrophilic properties. PVA, cellulose and Cts are polar polymers; thus a nonocomposite of PVA, Cts and NCC is likely to produce a material including excellent mechanical properties. Embedding NCC into PVA matrix changes the physicochemical properties of PVA and improves polymer structure in both molecular and morphological levels [23] . We have chosen Cts because it is one of the strongest biopolymers containing reactive amino groups. Moreover, this biopolymer has several important properties such as biocompatibility and anti-bacterial properties, which made it a suitable material for biomedical applications such as drug delivery, tissue engineering, wound healing and various antimicrobial strategies [24] . Also, Cts films have been used for food packaging, mainly due to their non-toxicity and biodegradability [25] . In this study we have prepared NCC from Whatman No. 1 filter paper via acid hydrolysis.

EXPERIMENTS

Materials

The materials used in this study, consist of Whatman No. 1 filter paper as a precourser for NCC, acetic acid (analytical purity), Sulfuric acid (98%) for acidic hydrolysis, PVA and low-molecular-weight Chitosan as matrices, which were obtained from Merk (Darmstadt, Germany) Company Inc.

Preparation of Nanocrystalline Cellulose

The NCC was obtained via acidic hydrolysis using traditional method. A Whatman No. 1 filter paper (98% α-cellulose, 80% crystallinity) was chopped and transferred into 70 ml of deionized water and blended by a (10 Speed Osterizer) Blender to turn a pulp. The pulp was filtered by a Buchner Funnel under vaccume filtration. Then 40 ml Sulfuric acid (98%) was added dropwise to the solution under vigorous stirring at 45°C until the slurry turned to a milky mixture for about 2 hours. After dilution with 200 ml of cold deionized water, the mixture was subjected to five washing/centrifugation cycles using a centrifuge apparatus (14,000 g, 15 min, Eppendorf, Germany). When pH was reached to about 4 - 5, the fine cellulose particles were dispersed into aqueous supernatant. The residue containing NCC, was then filtered.

Preparation of Cts/PVA/NCC Nanocomposite

Cts/PVA/NCC was prepared using traditional method. 1 g Cts was dissolved in 100 ml acetic acid (0.1 M) and stirred for 6 hours at room temperature. Then the solution was filtered in order to remove undissolved materials. PVA was dissolved in 20 ml of hot distilled water under constant stirring and then added to Cts solution followed by stirring for 2 hours. Various weight ratios of NCC (0, 5, 10 and 15% w/w) were added directly into Cts/PVA solution and then stirred for 24 hours. The mixtures were casted into petri dishes and placed in an air- convection oven at 60°C for 12 hours. The average thickness of every prepared film was about 1 mm.

CHARACTERIZATION METHODS

Fourier-Transform Infrared (FTIR) Spectroscopy

FT-IR spectroscopy was used to confirm that a reaction was occurred between PVA, Cts and NCC. The IR spectra of PVA/Cts/NCC films were recorded using a Bruker Optics Ft Tensor 27, Germany Instrument. FTIR spectra of cellulosic samples were recorded in the range of 400 - 4000 cm^{-1} by the resolution of 4 cm^{-1}. The samples were grounded into powder, mixed with KBr and then converted into thin pellets.

X-Ray Diffraction (XRD) Analysis

An X-ray diffractometer (X Pert Prompd, Phillips, Netherlands) was used to investigate the crystallinity of nanocellulose. Milled sample powder was analyzed using a Cu-K_α (λ = 1.54 angstrom) analyser. The angle of incidence was varied from 4° to 80° by the step of 0.02° and at the rate of 3°/min at 40 kV and 30 mA. The Segal method was used to calculate the sample crystallinity, using the height of the 2 0 0 peak (I_{200}, $2\theta = 22.93°$) and the lowest height between the 2 0 0 and 1 1 0 peaks (I_{AM}, $2\theta = 18°$). I_{200} represents both crystalline and amorphous regions while I_{AM} represents amorphous material only.

Scanning Electron Microscopy (SEM) Analysis

Scanning electron microscopy (SEM) photographs of the samples were captured using (Hitachi S4160, Germany) electron microscope. The samples were coated with gold using sputtering technique. Scanning electron microscopy (SEM) was used to investigate the morphology of different types of films.

Gel Swelling Property

First the films were dried at 40°C for 12 hours in an incubator and then weighed. Dried films were then immersed in distilled water for 1, 2, 4 and 8 hours. Wet weight of the films was measured after removing them from water and obliterating the surface adsorbed water. Then films were weighed immediately [26] . Gel swelling property of films was calculated by the following equation:

$$S = \left[\frac{w_s - w_d}{w_d} \right] \times 100$$

in which, S is the percentage of water adsorption of films at equilibrium state, w_s and w_d are the weights (in grams) of samples in swollen and dried states, respectively.

Biodegradation Property

Biodegradation property of different PVA/Cts/NCC films was determined by exposing the samples to compost mud. In this study, the degradation of films was evaluated by measurement of their weight loss, which refers to erosion of the molecules from solid phase to aqueous phase. The dissolved components were easily degraded by micro organisms in a natural environment.

RESULTS AND DISCUSSION

FTIR Spectroscopy Analysis

FT-IR spectra of NCC and the nanocomposites were shown in Figure 1. The spectrum of NCC was exhibited a broad band in the region of 3500 - 3200 cm^{-1}, which indicated the inter hydrogen bonded O-H stretching vibration of the OH groups in cellulose molecule. The spectra of all other samples were shown characteristic C-H stretching vibration around 2900 cm^{-1} for C-H stretching and the peak in the region of about 1641 cm^{-1} which was attributed to glycosidic C-O bonds. Besides, the vibrational peak at 1365 cm^{-1} in the samples B, C, D and E was related to the bending vibration of C-H bonds in aromatic ring. The band around 1423 cm^{-1} was attributed to the aliphatic C-H stretching of PVA. Furthermore, the peaks around 1430 cm^{-1} were associated with the C-H in plane bending of NCC.

X-Ray Diffraction Studies

The crystallinity of NCC was calculated to be 81.5% from Segal Method. This value was in the range of published values 86.7% for commercial microfibrillated cellulose (MFC). Pure Cts was shown two peaks at 2θ of 9.37°, 19.56° and a broad peak appearing at 2θ values in the range 19° - 28° was generally pertinent to the polymeric PANi chains. The peaks at 2θ values of 7.12° for MMT and 6.09° for MMT/Cts were also observed. As shown in Figure 2, all the four diffractograms display two well-defined peaks around 2θ of 12.5° (for 1 1 0 plane) and 2θ of 22.5° (for 2 0 0 plane) characteristic of cellulose [4] . Proper control of the hydrolysis conditions were resulted in selective degradation and preferential removal of amorphous cellulosic fraction. However, the average crystallite size was obtained from X-ray diffraction data using Scherrer's formula:

$$D = \frac{K\lambda}{\beta Cos\theta}$$

where k = 0.94, λ = 0.154056 nm and β is the full width at half maximum (FWHM) in radians. The prepared NCC had rectangular shape with average dimensions of 15.7 nm. The above experimental observations were indicated that NCC belonged to semi-crystalline polymer, which were contained crystalline and amorphous regions. Accordingly, the above results were demonstrated that the hydrolysis was taken place in the amorphous region. This increase of crystallinity after acidic treatment has been reported by several authors [27] [28] .

Scanning Electron Microscopy (SEM) Analysis

The morphology of Cts/PVA/NCC nanocomposites was assigned using SEM. Scanning electron micrographs of the nanocomposite films with different contents of NCC were shown in Figure 3. The random orientation and good dispersion of NCC in matrices was observed. These results were indicated an excellent compatibility between the three components of every nanocomposite. According to the SEM results, it was obvious that the whiskers of NCC in nanocomposites were rod like. The morphology of NCC was appeared unchanged after blending with Cts/PVA matrix.

Gel Swelling Property Analysis

Figure 4 was illustrated the effect of the concentration of NCC on swelling percentage and water uptake of Cts/ PVA films. The presence of NCC in the films was reduced the swelling percentage of Cts/PVA films. Water uptake of nanocomposites was depended on the nature of matrix and filler. The decreased water uptake could be ascribed to the high crystallization of NCC which was less hydrophilic than Cts/PVA composite and the formation of strong filler-matrix interactions [26] [29] [30] . NCC was acted as an interpenetrating network within the matrix and prevented the swelling of the Cts/PVA films when taken into the water [31] .

Biodegradation Analysis

The weight loss in biodegradation of Cts/PVA and Cts/PVA/NCC films was shown in Table 1. Accordingly, it was observed that the degradation occurred in a faster rate in the presence of NCC in the Cts/PVA matrix. Consequently, the nanocomposites could be easily degraded in natural condition.

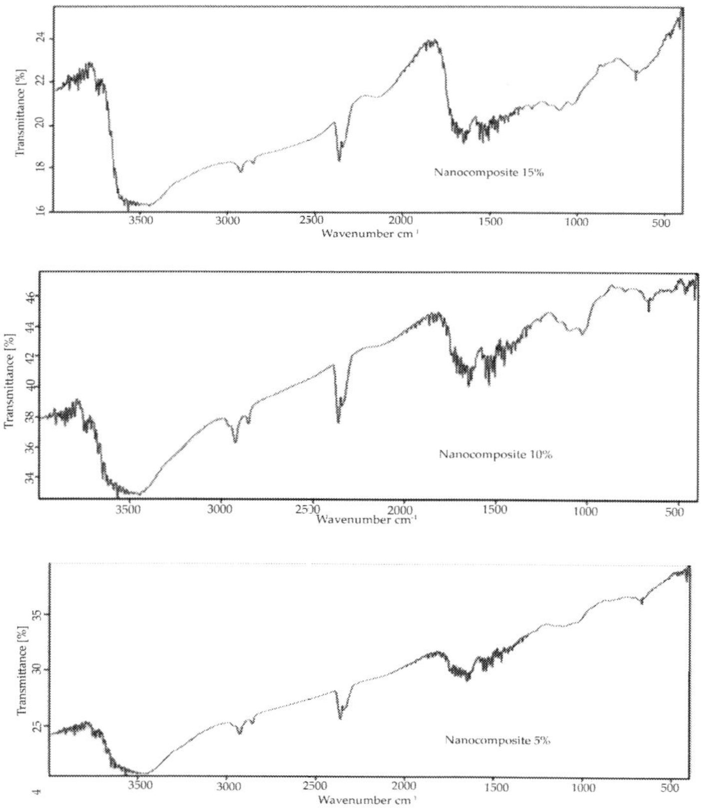

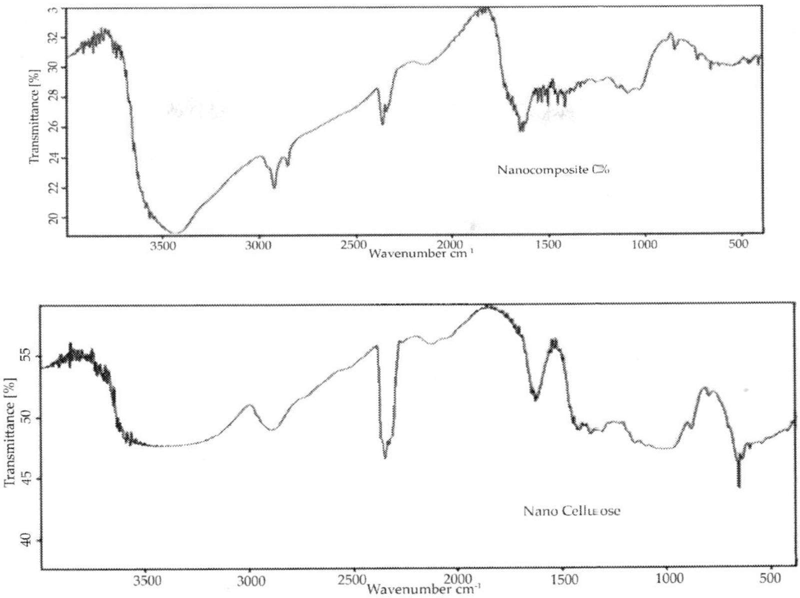

Figure 1: FT-IR spectra of NCC and nanocomposites with different NCC concentrations.

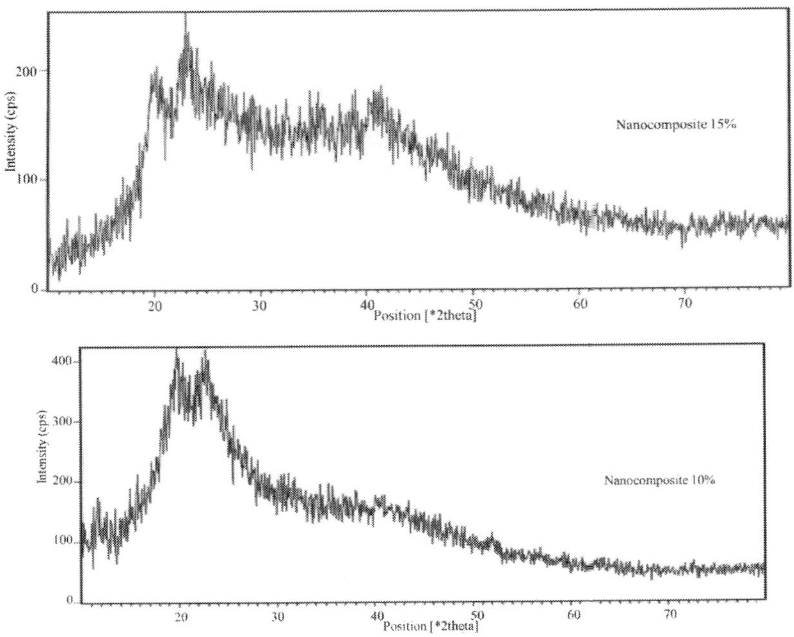

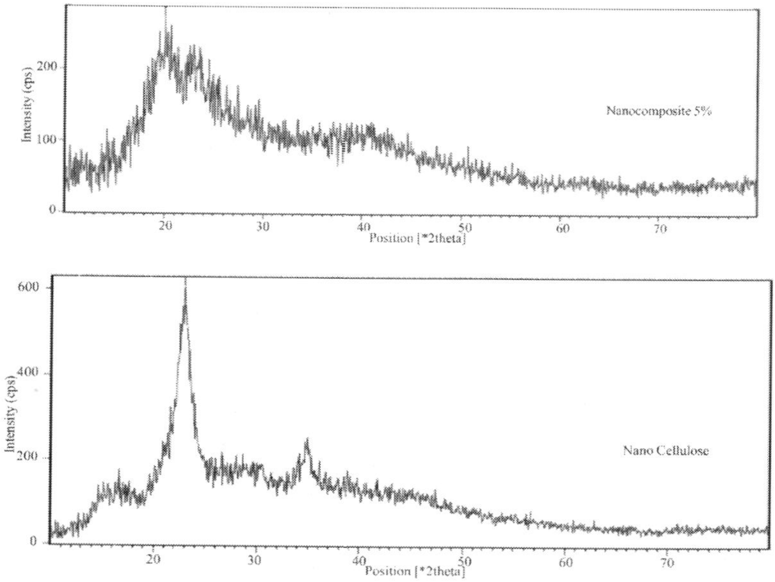

Figure 2: XRD patterns of NCC and nanocomposites with different NCC.

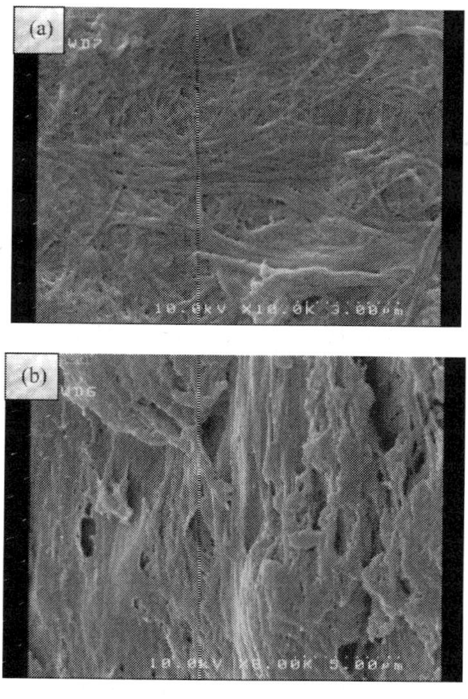

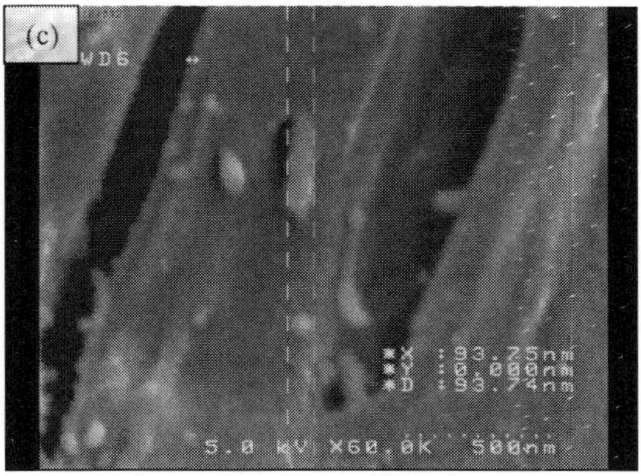

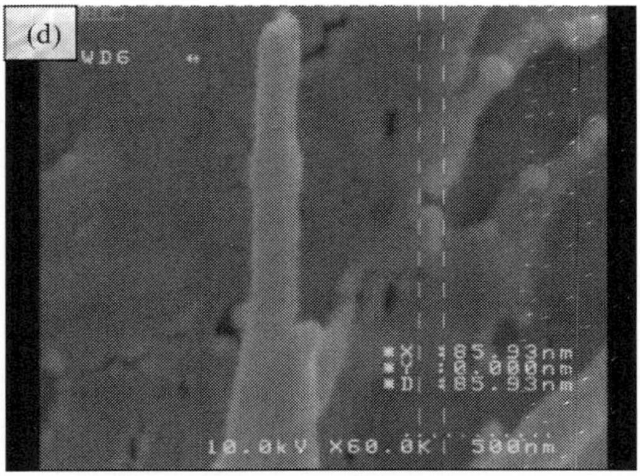

Figure 3: FESEM pictures of (a) NCC, (b) nanocomposite 5%, (c) nanocomposite10% and (d) nanocomposite 15%.

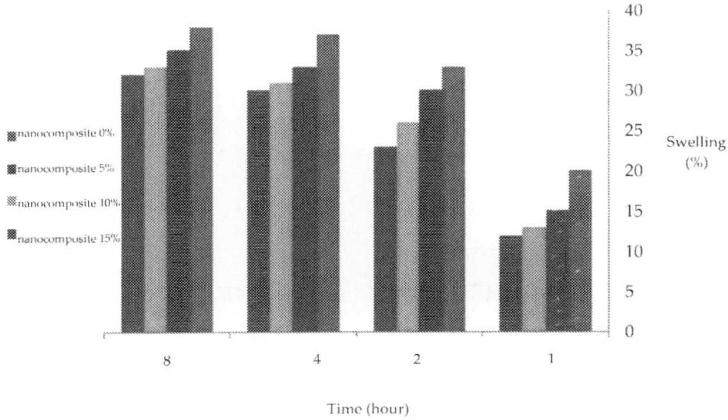

Figure 4: The effect of NCC content in the gel swelling property of Cts/PVA films.

Table 1: Biodegradation for the Cts/PVA and Cts/PVA/NCC films

Films	Weight loss (g)			
	1st day	2nd day	4th day	5th day
Nanocomposite 0%	0.16	0.12	0.11	0.09
Nanocomposite 5%	0.21	0.23	0.13	0.07
Nanocomposite 10%	0.32	0.30	0.15	0.03
Nanocomposite 15%	0.40	0.33	0.19	0.02

CONCLUSION

In this research, NCC was synthesized from Whatmann No. 1 filter paper. Cts/PVA/NCC nanocomposites were prepared in different concentrations of NCC [5% (B), 10% (C) and 15% (D)]. Further studies should be focused on the investigation of other potential applications of NCC and its nanocomposites in combination with other polymers. NCCs which were isolated by sulfuric acid hydrolysis, were rod-like whiskers with an average size of about 15.7 nm observed by XRD analysis, and the crystallinity of NCC was about 81.5%. SEM images of NCC showed the dimensions of about 85 nm wide. NCC improved the barrier properties of Cts/PVA composite films via reduction of swelling property. Although, a comparison with related literature results showed that the nanocomposites which were investigated in this work were amongst the best materials reported so far, this approach appeared really promising for the design of a wide range of new related nanocomposites. Surface morphology of nanocomposite films revealed a homogeneous dispersion of NCC into the Cts/PVA matrix. The prepared Cts/ PVA/NCC films were biodegradable.

ACKNOWLEDGEMENTS

This research has been accomplished under financial support of the Department of Chemistry, Faculty of Science, and cordially laboratorial collaboration of our colleagues at the Department of Agronomy, Faculty of Agriculture, University of Zabol, which should be appreciated.

REFERENCES

1. Whitesides, G.M. (2005) Nanoscience, Nanotechnology, and Chemistry. Nanoscience and Chemistry, 1, 172-179. http://dx.doi.org/10.1002/smll.200400130
2. Kester, J.J. and Fennema, O.R. (1986) Edible Films and Coatings: A Review. Food Technology, 40, 47-59.
3. Gennadios, A. (2002) Protein-Based Films and Coatings. CRC Press, Boca Raton, 367-392. http://dx.doi.org/10.1201/9781420031980
4. Klemm, D., Heublein, B., Fink, H.P. and Bohn, A. (2005) Cellulose: Fascinating Biopolymer and Sustainable Raw Material. Angewandte Chemie International Edition, 44, 3358-3393. http://dx.doi.org/10.1002/anie.200460587 [Citation Time(s):2]
5. Zimmerman, T., Pohler, E. and Schwaller, P. (2005) Mechanical and Morphological Properties of Cellulose Fibril Reinforced Nanocomposites. Advanced Engineering Materials, 12, 1156-1161. http://dx.doi.org/10.1002/adem.200500157
6. Azizi-Samir, M.A.S., Alloin, F. and Dufresne, A. (2005) Review of Recent Research into Cellulosic Whiskers, Their Properties and Their Application in Nanocomposite Field. Biomacromolecules, 6, 612-626. http://dx.doi.org/10.1021/bm0493685
7. Habibi, Y., Lucia, L.A. and Rojas, O.J. (2010) Cellulose Nanocrystals: Chemistry, Self-Assembly, and Applications. Chemical Reviews, 110, 3479-3500. http://dx.doi.org/10.1021/cr900339w
8. Cao, X.D., Habibi, Y., Magalhaes, W.L.E., Rojas, O.J. and Lucia, L.A. (2011) Cellulose Nanocrystals-Based Nanocomposites: Fruits of a Novel Biomass Research and Teaching Platform. Current Science, 100, 1172-1176.
9. Siro, I. and Plackett, D. (2010) Microfibrillated Cellulose and New Nanocomposite Materials: A Review. Cellulose, 17, 459-494. http://dx.doi.org/10.1007/s10570-010-9405-y
10. Liang, S.M., Zhang, L.N., Li, Y.F. and Xu, J. (2007) Fabrication and Properties of Cellulose Hydrated Membrane with Unique Structure. Macromolecular Chemistry and Physics, 208, 594-602. http://dx.doi.org/10.1002/macp.200600579
11. Wu, Y.T., Zhou, Z., Fan, Q.Q., Chen, L. and Zhu, M.F. (2009) Facile

In-Situ Fabrication of Novel Organic Nanoparticle Hydrogels with Excellent Mechanical Properties. Journal of Materials Chemistry, 19, 7340-7346. http://dx.doi.org/10.1039/b909125d

12. Zhang, Y., Yu, J.R., Zhou, C.J., Chen, L. and Hu, Z.M. (2010) Preparation, Morphology, and Adhesive and Mechanical Properties of Ultrahigh-Molecular-Weight Polyethylene/SiO2 Nanocomposite Fibers. Polymer Composites, 31, 684-690.
13. Zhou, C.J., Wang, S.F., Zhang, Y., Zhuang, Q.X. and Han, Z.W. (2008) In Situ Preparation and Continuous Fiber Spinning of Poly(p-phenylenebenzobisoxazole) Composites with Oligo-Hydroxyamide-Functionalized Multi-Walled Carbon Nanotubes. Polymer, 49, 2520-2530. http://dx.doi.org/10.1016/j.polymer.2008.04.003
14. Pu, Y.Q., Zhang, J.G., Elder, T., Deng, Y., Gatenholm, P. and Ragauskas, A. (2007) Investigation into Nanocellulosics versus Acacia Reinforced Acrylic Films. Composites Part B: Engineering, 38, 360-366. http://dx.doi.org/10.1016/j.compositesb.2006.07.008
15. Zhang, J., Jiang, N., Dang, Z., Elder, T.J. and Ragauska, A.J. (2008) Oxidation and Sulfonation of Cellulosics. Cellulose, 15, 489-496. http://dx.doi.org/10.1007/s10570-007-9193-1
16. Lima, M.M.D. and Borsali, R. (2004) Rod-Like Cellulose Microcrystals: Structure, Properties and Applications. Macromolecular Rapid Communications, 25, 771-787. http://dx.doi.org/10.1002/marc.200300268
17. Dufresne, A. (2006) Comparing the Mechanical Properties of High Performances Polymer Nanocomposites from Biological Sources. Journal of Nanoscience and Nanotechnology, 6, 322-330.
18. Dufresne, A. (2008) Polysaccharide Nanocrystal Reinforced Nanocomposites. Canadian Journal of Chemistry, 86, 484-494. http://dx.doi.org/10.1139/v07-152
19. Teixeira, E.M., Pasquini, D., Curvelo, A.A.S., Corradini, E., Belgacem, M.N. and Dufresne, A. (2009) Cassava Bagasse Cellulose Nanofibrils Reinforced Thermoplastic Cassava Starch. Carbohydrate Polymers, 78, 422-431. http://dx.doi.org/10.1016/j.carbpol.2009.04.034
20. Kim, J.T. and Netravali, A.N. (2010) Mechanical, Thermal, and

Interfacial Properties of Green Composites with Ramie Fiber and Soy Resins. Journal of Agricultural and Food Chemistry, 58, 5400-5407. http://dx.doi.org/10.1021/jf100317y

21. Khalil, H.P.S.A., Bhat, A.H. and Yusra, A.F.I. (2012) Green Composites from Sustainable Cellulose Nanofibrils: A Review. Carbohydrate Polymers, 87, 963-979. http://dx.doi.org/10.1016/j.carbpol.2011.08.078
22. Masuhiro, T., Giuliano, F. and John, S.C. (1994) Structure and Compatibility of Poly(vinyl alcohol)-Silk Fibroin (PVA/SA) Blend Films. Journal of Polymer Science B: Polymer Physics, 32, 243-248. http://dx.doi.org/10.1002/polb.1994.090320205
23. Siddaramaiah, R.T., Ravi, P., Kumar, K.P. and Jagadeesh, K.S. (1999) Effect of Fillers on the Properties of LDPE Films. Journal of Polymer Materials, 16, 233-237.
24. Dash, M., Chiellini, F., Ottenbrite, R.M. and Chiellini, E. (2011) Chitosan-A Versatile Semi-Synthetic Polymer in Biomedical Applications. Progress in Polymer Science, 36, 981-1014. http://dx.doi.org/10.1016/j.progpolymsci.2011.02.001
25. De Azeredo, H.M.C. (2009) Nanocomposites for Food Packaging Applications. Food Research International, 42, 1240-1253. http://dx.doi.org/10.1016/j.foodres.2009.03.019
26. Jin, J., Song, M. and Hourston, D.J. (2004) Novel Chitosan-Based Films Cross-Linked by Genipin with Improved Physical Properties. Biomacromolecules, 5, 162-168. http://dx.doi.org/10.1021/bm034286m
27. Azizi-Samir, M.A.S., Alloin, F., Paillet, M. and Dufresne, A. (2004) Tangling Effect in Fibrillated Cellulose Reinforced Nanocomposites. Macromolecules, 37, 4313-4316. http://dx.doi.org/10.1021/ma035939u
28. Tang, L.G., Hon, D.N.S., Pan, S.H., Zhu, Y.Q., Wang, Z. and Wang, Z.Z. (1996) Evaluation of Microcrystalline Cellulose. I. Changes in Ultrastructural Characteristics during Preliminary Acid Hydrolysis. Journal of Applied Polymer Science, 59, 483-488. http://dx.doi.org/10.1002/(SICI)1097-4628(19960118)59:3<483::AID-APP13>3.0.CO;2-V
29. Dufresne, A., Dupeyre, D. and Vignon, M.R. (2000) Cellulose

Microfibrils from Potato Tuber Cells: Processing and Characterization of Starch-Cellulose Microfibril Composites. Journal of Applied Polymer Science, 76, 2080-2092. http://dx.doi.org/10.1002/(SICI)1097-4628(20000628)76:14<2080::AID-APP12>3.0.CO;2-U

30. Li, Q., Zhou, J. and Zhang, L. (2009) Structure and Properties of the Nanocomposite Films of Chitosan Reinforced with Cellulose Whiskers. Journal of Polymer Science Part B: Polymer Physics, 47, 1069-1077. http://dx.doi.org/10.1002/polb.21711

31. Svagan, A.J., Hedenqvist, M.S. and Berglund, L. (2009) Reduced Water Vapour Sorption in Cellulose Nanocomposites with Starch Matrix. Composites Science and Technology, 69, 500-506. http://dx.doi.org/10.1016/j.compscitech.2008.11.016

13
CHAPTER

IN VITRO PROPERTIES OF CHITOSAN NANOPARTICLES INDUCE APOPTOSIS IN HUMAN LYMPHOMA SUDHL-4 CELL LINE

Umesh K. Parida[1], Niranjan Rout[2], Birendra K. Bindhani[1]

[1] School of Biotechnology, KIIT University, Bhubaneswar, India

[2]Acharya Harihar Regional Cancer Center, Cuttack, India

Keywords: Chitosan Nanoparticles, Sodium Triphosphate, SUDHL-4, Apoptosis, MTT Assay

ABSTRACT

In this study, the possible mechanisms were investigated with chitosan nanoparticles using sodium tripoly-phosphate and effects on human lymphoma SUDHL-4 in vitro. It was characterized by XRD, FTIR, TGA, particle Size, zeta potential, SEM & TEM. Different techniques such as cell proliferation, ultra structure changes, DNA fragmentation, phase distribution of cell cycle, MTT assay, MMP, agarose gel electrophoresis of DNA, flow cytometry and electron microscopy were used with treatment of different concentrations of CH-NPs (25, 50, 75, 100 μg/ml) at different time periods. Electron microscopy study revealed that the chitosan nanoparticles showed 78 nm particle size which is a high surface charge as 52 mV. Inhibition of chitosan nanoparticles after 48h treatment was marked in cell proliferation of SUDHL- 4 with an IC_{50} value of 5 μg/ml. Electron microscopy showed typical necrotic cell morphology after treatment of chitosan nanoparticles. The DNA degradation related with necrosis was determined using agarose electrophoresis and loss of MMP & occurrence of apoptosis was analyzed by flow cytometry. Chitosan nanoparticles with low molecular weight (LMW) were comparatively stable in medium containing aqueous and rate of dissolution was slow in acidic medium. Results of this present study clearly provided information that the chitosan nanoparticles effectively inhibit the proliferation of SUDHL-4 through multiple mechanisms in vitro and this novel formulation can open a new avenue against human Lymphoma.

INTRODUCTION

Biomedical applications of chitosan include, drug delivery, tissue engineering, wound dressings, implant coatings and therapeutic agent delivery systems. Chitosans are biocompatible, profuse, nontoxic, biodegradable and renewable carbohydrate polymers [1-7].

Recent research has revealed the antimicrobial potential of chitosan on its M_r and DA [8-12] and higher viscosity of M_r and chitosan solutions have its multidimensional benefit. Alternatively, the low molecular weight chitosans could be depolymerised by physical, acidic or enzymatic methods. The highest bacterial activity towards pathogenic bacteria was found in LMWC (5 - 10 kDa) [13], whereas, 20 kDa product prevented progression of diabetes mellitus and showed a higher affinity for lipopolysaccharides (LPS) than the native chitosan ~140 kDa [14]. The sensible use of LMWC in milk safeguarding and oral hygiene is also reported [15].

It was reported that chitosan is biocompatibility and biodegradability [16,17] applied in drug delivery systems to prepare encapsulation of drugs, enzymes, proteins and DNA in nanospheres or microspheres level

[18-20]. Drugs for biological activities could be developed by using chitisan only. Both soluble and microspheres chitosan showed a type of toxic in murine melanoma cell line and chitosan with hydrochloride exhibited extra toxic with an IC_{50} [21]. Dose dependent inhibitory effects of amino cationic chitosan derivatives showed the proliferation of numerous tumour cell lines with lowest IC_{50} towards liver cancer [22,23].

The most important properties of CH-NPs could provide a high resemblance used for negatively charged biological membranes and targeting specific site in vivo [13]. The CH-NPs with different concentrations could inhibit the dissimilar proliferated tumour cell lines whereas it could be toxic against normal liver cells [19]. The present investigations illustrate the possible mechanisms of CH-NPs and effects on the proliferation of human Lymphoma, SUDHL-4 cell line. It was characterized using electron microscopy, MTT assay, agarose electrophoresis of DNA, and flow cytometry.

MATERIALS AND METHODS

Chitosan was obtained from himedia and sodium tripolyphosphate (TPP), NaOH, acetic acid, NaH_2PO_4, were obtained from Hindustan scientific Linker, Odisha, India. SUDHL-4 (Cell Lines) was maintained in School of Biotechnology, KIIT University, Odisha, India. It was grown in medium (i.e., RPMI-1640) supplemented with 10% foetal calf serum in a humidified incubator at 37°C along with 5% CO_2. Collagen coated culture flasks were used when necessary and water was always used in demineralised form.

Isolation of LMWC

Different concentration of chitosan solution was treated with pronase in the ratio 100:1 (w/w), incubated for different periods at 37°C followed by arresting the reaction by heat denaturing the enzyme (100°C) and added equal volume of 2 N NaOH. The precipitated LMWC was obtained after centrifugation (3000 rpm, 10 min). Then, it was dialyzed against deionised water using a membrane and freeze dried.

Preparation of Nanoparticles

There are three different types of chitosan such as 6 kDa, 8 kDa and 10 kDa. The chitosan with 6 kDa was used more preferably as it is more compressed then the chitosan 8 kDa and 10 kDa. Chitosan nanoparticles (CHNPs) were synthesized via the ionotropic gelation [5] of chitosan with TPP anions. The synthesis of different concentrations of CH-NPs was first optimized and then dissolved in acetic acid (1%). The sodium tripolyphosphate (TPP) solution (1 mg/ml) was prepared by double distilled water. Chitosan nanoparticles were spontaneously fabricated with the drop wise addition of TPP solution (CH/TPP = 4:1) under magnetic stirring (1000 rpm) at room temperature (Figure 1(a)).

The opalescent suspension was shaped under the similar above mentioned conditions. The nanoparticles were separated by centrifugation at 20,000 g and 14°C for 30 minutes and supernatants was discarded, freezedried and stored at 5°C ± 3°C. The weight of freeze-dried nanoparticles was calculated.

Characterization

LMW Chitosan

Three dissimilar LMW chitosan were prepared on the basis of H NMR technique and viscosity method respectively.

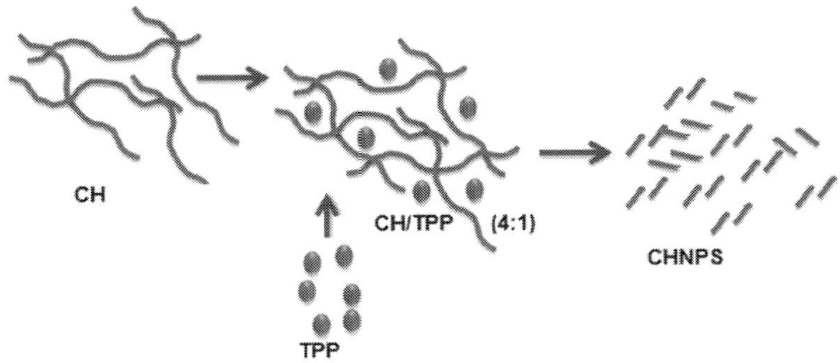

(a)

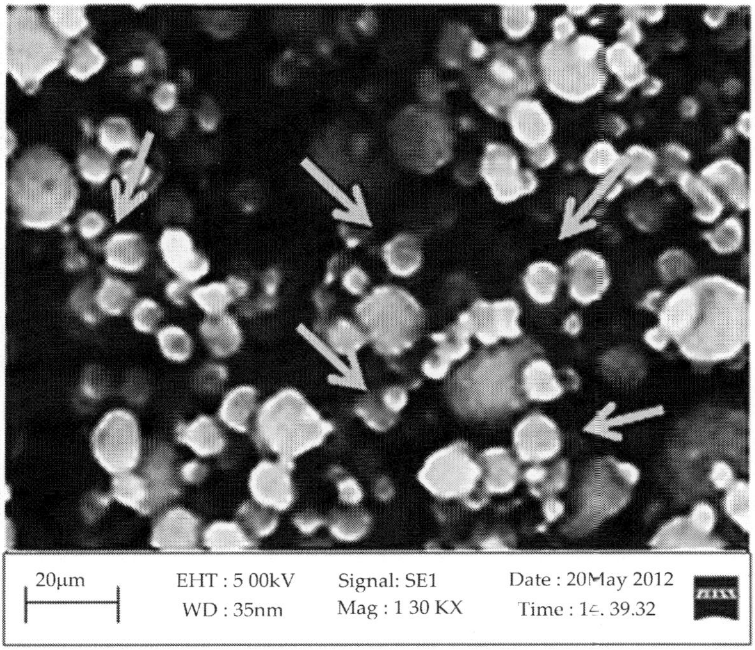

(b)

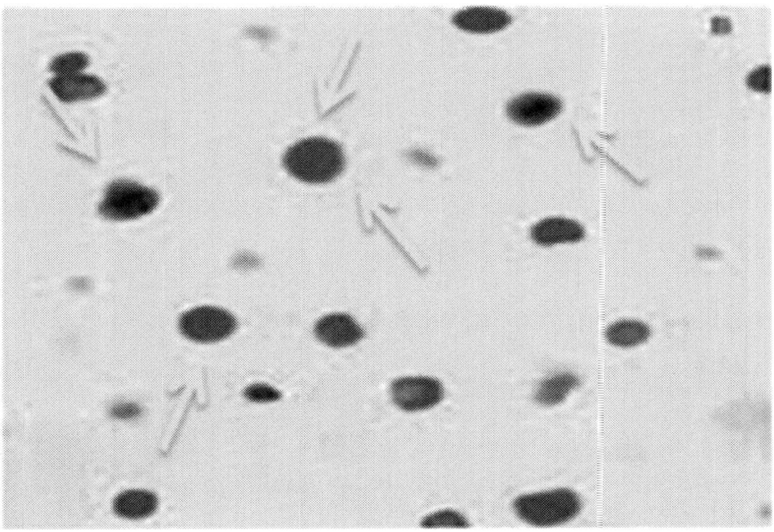

(c)

Figure 1: (a) A schematic design for preparation of the chitosan nanoparticles using sodium tripolyphosphate (TPP); (b) Scanning electron microscopy of chitosan nanoparticles; (c) Transmission electron microscopy of chitosan nanoparticles.

The M_r of chitosan used in this study was 6, 8 and 10 kDa. All chitosan grades were found to be entirely deacetylated.

SEM (Scanning Electron Microscopy) Analysis

To observe the morphological features, Field Emission Scanning Electron Microscopy, JEOL, JSM-6700F, Japan was used. By using sputter coated with a 5 nm thick gold, the samples were sprinkled.

TEM (Transmission Electron Microscopy) Analysis

TEM analysis was performed on a Hitachi H-8100 electron microscope with an acceleration voltage of 200 kV.

XRD (X-Ray Diffraction) Analysis

It was carried out using BEDE D-3 system with Cu Kα radiation at a generator or voltage of 40 kV and a generator current of 100 mA. The samples were scanned from scanned from $2\theta = 1 - 100°$ at a scanning rate of 2°/min.

Zeta Size and Potential Analysis

A Malvern Instrument, MAL 1037088, USA was used to determine the standard size of particle and zeta potential of nanoparticles. Disposable zeta cells measurement was carried with ultra pure water at 25°C with a −50 mV latex standard calibrated frequently. The mean zeta potential was carried out using phase analysis light scattering technique.

FT-IR Analysis

A Perkin-Elmer Model of FTIR spectrophotometer, USA within the range of 4000 - 400 cm^{-1} was used for the samples analysis. Approximately, sample of 5 mg was with KBr (100 mg) and condensed into pellet using hydraulic press. The KBr pellet methods were used for all FT-IR spectra analysis.

Thermo Gravimetric Analysis (TGA)

Thermo gravimetric analysis (Perkin Elmer Model, USA) was used having unit of microprocessor temperature control with TA data station. The samples mass generally in the range of 2 - 3 mg was used. An equipment consist of both sample pan with balance system was placed in temperature ranges from 25°C to 800°C with 50 cm^3/min. flow rate of nitrogen. The mass of sample pan was recorded constantly as a function of temperature.

Cell Viability Assay

In this case, cell (SUDHL-4) suspension with 100 μl aliquot containing 10^6 cells was added to every well of a 96 well plate (Corning, USA) and cultured 24 h. Then cells were treated immediately with different concentrations of CH-NPs such as 25 μg/ml, 50 μg/ml, 75 μg/ml and 100 μg/ml for another 24 or 48 h. The tetrazolium dye assay was used to know the consequence of diverse treatments on cell viability. The data in this study were expressed as mean ± SD.

DNA Fragmentation

First cells were treated with 100 μg/ml CH-NPs for 6 or 24 h, and then collected cells were washed using PBS followed by lyses. Then, such cells were used for extraction of DNA with phenol/chloroform/isoamyl alcohol (25:24:1) followed by chloroform. Precipitation of DNA was observed with two volumes of ethanol followed by 0.3 mol/l sodium acetate. Finally, DNA samples were obtained and visualized by agarose gel electrophoresis.

Determination of Mitochondrial Membrane Potential (MMP)

In this study, SUDHL-4 cells were treated with different concentrations (25 μg/ml, 50 μg/ml, 75 μg/ml and 100 μg/ml) of CH-NPs and followed by staining using rhodamine 123 (10 μg/ml) which is simply sequestered by MMP [13]. The rhodamine 123 was washed of the cells after loss of MMP. The mitochondrial membrane potential was find out using FACS calibur flow cytometer. For MMP analysis, the Cell Quest software program (BD Phar Mingen, Franklin Lakes, USA) was used.

Cell Cycle Analysis

Determination of DNA content and peaks of the apoptotic cells was done using flow cytometry. SUDHL-4 cells were grown in RPMI-1640 supplemented with 10% FBS. Different concentrations (25 μg/ml, 50 μg/ml, 75 μg/ml and 100 μg/ml) of CH-NPs were treated for 24h, after treatments cells were collected and stored at 4°C. Fixation cells were washed, centrifuged and re-suspended in propidium iodine (0.05 mg/ml), RNase (100 U/ml) in PBS. The sample was incubated and analysed by Calibur flow Cytometer. The Cell Quest software and MOdFit software was used to determine the cell cycle data. In this study the negative controls were also maintained against the positive controls.

RESULTS AND DISCUSSIONS

Determination of LMWC

The determination of M_r (i.e., 71 ± 2 kDa) of native chitosan was done using GPC on Sepharose CL-4B and calculated its value in accordance with viscometry. In good agreement with each other, the DA was calculated from IR and ^{1}H-NMR data. Different molecular weight (i.e., 10.5 and 6.5 kDa) values were calculated in different time intervals (i.e., 2 - 8 h) by HPLC and GPC. LMWC of free-NH_2groups and the column material showed minimum interaction using acetate buffer. The preparations of LMWC showed indicative molecular homogeneity with single symmetrical peak (Table 1) [24- 28].

Scanning Electron Microscopy and Transmission Electron Microscopy Analysis

The bare chitosan nanoparticle (CH-NP) with size magnifications was done by SEM analysis (Figure 1(b)).

Table 1: Characteristics and percent yield of LMWC

Chitosan	Mr (kDa)	DA (%)		Yield %
		IR	H-NMR	
Native	71.0 ± 2.0[a]	26.78	24.53	-
LMWC—2 h	10.50 ± 0.15[b]	18.04	17.87	82
LMWC—4 h	8.10 ± 0.12[b]	15.34	15.08	78
LMWC—8 h	6.50 ± 0.12[b]	13.29	13.34	76

[a]Determined by GPC on sepharose CL-4B column; [b]Determined by HPLC.

The morphology of bare chitosan nanoparticle was shown about 78 nm in diameter with spherical in shape and found to be agreement. The particle size was decreased due to its aggregation and high specific surface energy [29].

In TEM analysis, chitosan nanoparticles exhibited round shaped due to vesicle disruption (Figure 1(c)) [30].

Stability and Biocompatible of the Nanoparticles

The biocompatibility nanoparticles with storage ability are great concern in pharmaceuitical fields. It is known that small size particles are inclined to agglomerate due to reduce free surface area and energy. Different concentrations of TPP were prepared with 2 - 90 days storage time at 4°C to determine the size of nanoparticles (Figure 2(a)). Each and every particle was stable with neglectable fluctuated size within 15 days and some particles were shown larger in size at 90 days. The best storage stability was observed in different concentration of TPP (1.0 - 1.25 mg/ml at pH 6.0). The nanoparticles were precipitated continuously in undisturbed condition for some days and re-dispersed easily by moderate shaking.

At higher concentrations of TPP, the chitosan nanoparticles showed higher degree of cross linking whereas it is acceptable in decrease concentration of chitosan. Thus, it is resulted more compact particle structure. At the same time, particles showed smaller net charge due to neutralization of amino groups. It was also observed that smaller size particles structure due to compact and weakened charge repulsion in higher concentration of TPP.

XRD Analysis

In XRD analysis both pure and cross linked chitosans have been studied in the 2θ range of 5° to 60°. The XRD profile of chitosan shows peaks at 23° and 50° 2θ (Figure 2(b)). The crystallinity of the chitosan (3.02%) in

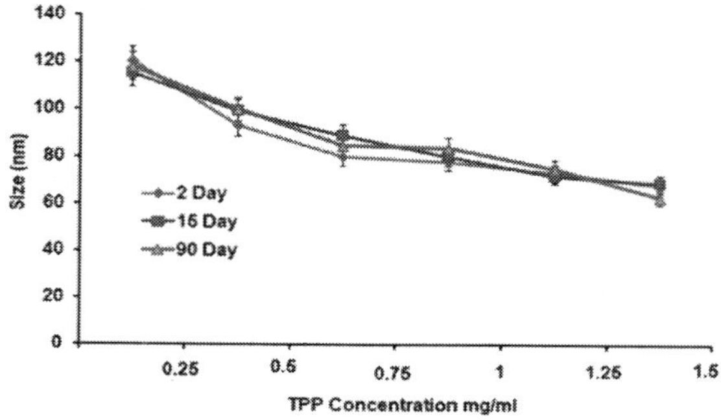

(a)

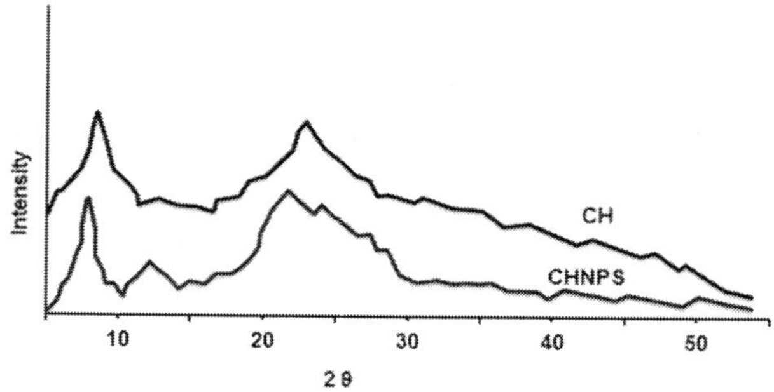

(b)

Figure 2: (a) Alteration of average size of chitosan nanoparticles with different TPP concentration during a storage time of 2 - 90 days at 4°C; (b) XRD patterns of chitosan and developed membranes.

creases with the increase in degree of deacetylation and is attributed to enlarge in intermolecular hydrogen bonding due to the occurrence of more free NH_2 groups (Higher Degree of Deacetylation) within the molecular structure, which in turn results the better packing of the macromolecular polymeric chains and consequent increase in the crystallinity.

Surface Morphology of SUDHL-4

The distraction of nuclear and cellular membrane exhibited necrosis under acute physiological stimuli condition. This important cause has been used to make a distinction necrosis from apoptosis of cellular membrane [31]. The chitosan nanoparticles treated with SUDHL-4 cells showed an ultra structure under SEM and TEM. Distinctive morphological characteristics of necrosis were observed in SUDHL-4 cells after treatment of chitosan nanoparticles. The SEM analysis showed both surface morphology and normal shape in control SUDHL-4 cells but large numbers of irregular microvilli were found in cell surface. After 30 minutes treatment of chitosan nanoparticles, cell death was induced due to leakage of early membrane and reduction of microvilli. After 2 h treatment of chitosan nanoparticles showed disintegration of microvilli and asymmetrical tiny holes with fracturing membrane solubilisation. The SUDHL-4 cells thus, damaged extensively and looked like as honeycomb shape after 4 h treatment. This type of necrotic cell death was observed due to pore forming surface morphology as well as loss of membrane integrity after the interface of CH-NP and plasma membrane. In TEM analysis, disruptions of cytoplasm and enlargement of organelles were observed in chitosan nanoparticles treated with necrotic SUDHL-4 cells whereas disrupted microvilli along with normal organelles in untreated cells. But vacuolated cells and complete disruption of plasma membrane were mainly observed after 24 h treatment with chitosan nanoparticles [32] (Figures 3(A)-(D)).

Zeta Size and Potential of Chitosan Nanoparticles

The normal chitosan nanoparticle size was 65 nm and average distribution size ranges starting 46 to 83 nm (Figure 4). In zeta potential, the stability of particle was significantly inclined on surface charge during the electrostatic repulsion among the particles. Our scanning electron microscopy studies of chitosan nanoparticles were almost coincide with the average particle size of chitosan nanoparticles (CH-NP). Measurement of zeta potential was one of the major indexes to determine the stability of nanoparticle suspension. Nanoparticles with

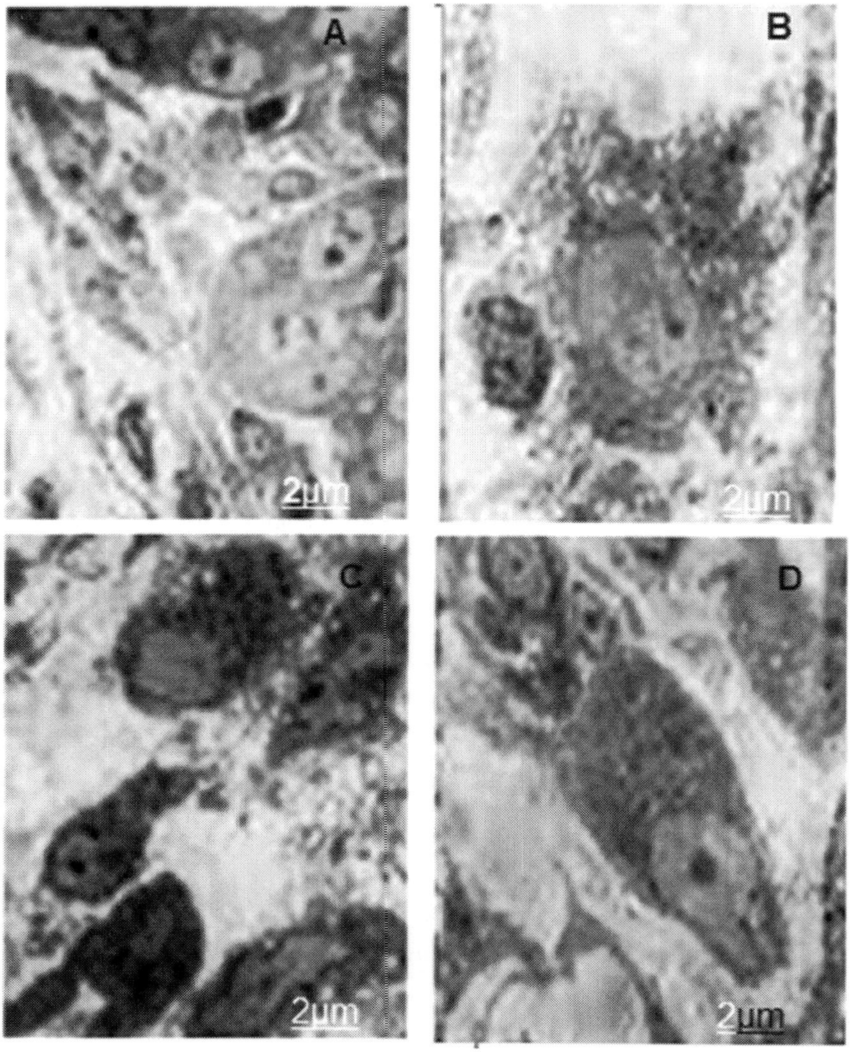

Figure 3: (A) Surface morphology of SUDHL-4 cells treated with 100 μg/mL chitosan nanoparticles for control cells; (B) 30 min; (C) 2 h; and (D) 4 h.

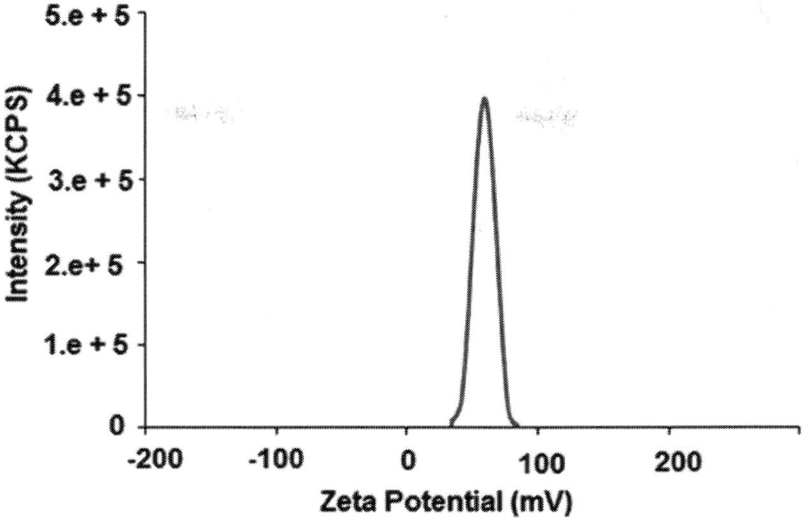

Figure 4: Size distributions of chitosan nanoparticles.

high electric surface charge showed high value of zeta potential due to repulsive forces between particles which leads to stop it from aggregation [33,34].

FTIR Analysis

In Figure 5(a), the transmission infrared spectra of powdered sodium tripolyphosphate (TPP), chitosan (CH) and ionically cross-linked chitosan/sodium tripolyphosphate (synthesized by cross-linking in TPP solution) are shown. A broad peak at 3433 cm^{-1} was shown in the bare $CH^{-}NP$ which corresponds to NH vibrational frequency. Generally, N-H stretching at 3500 - 3200 nm and N-H bend at 1550 - 1450 nm are observed in primary amino functional group. In FTIR spectrum, the region 4000 - 2000 nm was excluded due to interference of water molecules. The FTIR of TPP has both symmetric stretching and also asymmetric bending at 1022 nm and at 615 nm respectively. In such region (i.e., 615 nm and 1022 nm), two peaks were observed due to presence of phosphate in TPP chitosan [35,36]. It was also depicted the ionic bonding among the TPP and primary amine groups in TPP chitosan. Therefore, it implies that the occurrence of ammonium groups in TPP chitosan for such crosslinking reaction [37].

TGA Analysis

TGA analysis is normally used to determine the sample stability and loss of weight in diverse temperature. In this present study, TGA analysis of chitosan nanoparticles showed decomposition patterns in four stages (Figure 5(b)). In first stage, loss of weight (i.e., 6%) at 175°C was observed because of the release of water molecules. In second stage, loss of weight (i.e., 22% approximately) at 365°C was observed due to chitosan decomposition. In third stage, the loss of weight (i.e., 19%) was observed at 500°C caused by decomposition. At last, in fourth stage, such decomposition of chitosan was observed at 635°C with 16.79% of weight loss. However, the loss of weight

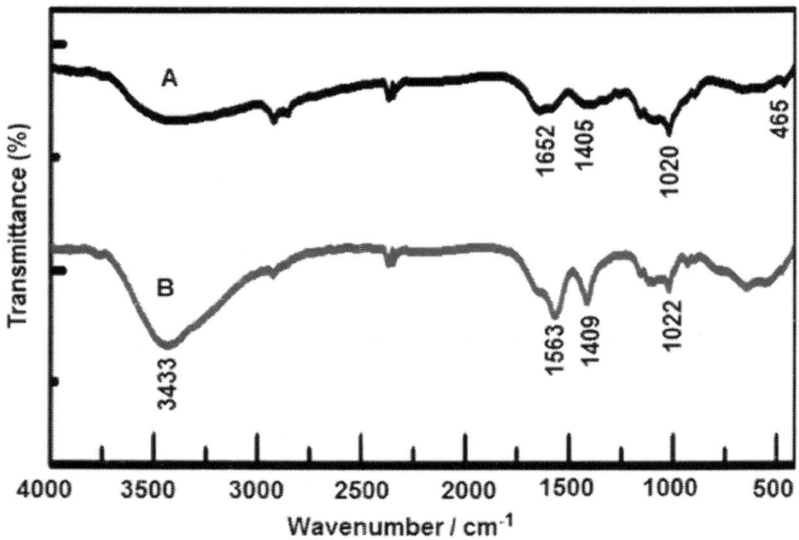

(a)

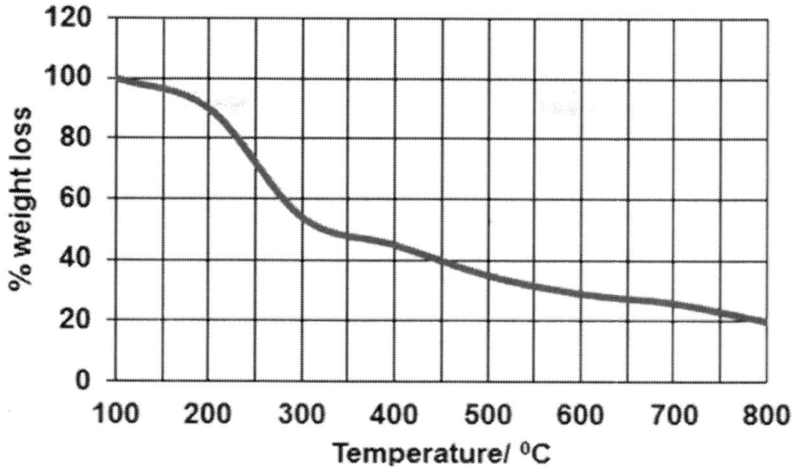

(b)

Figure 5: (a) FT-IR spectrum of CS, bare CS-NP; (b) Thermogravimetric analysis of CS-NP.

was also examined in two stages due to decomposition of chitosan within the range of 320°C and 470°C [38,39]. Our result could correlate with this report.

Cell Viability Assay

The MTT cell viability assay of SUDHL-4 cells treated using different concentrations of CH-NPs ranging as of 25 to 100 μg/ml were observed. The viability cells were inhibited by CH-NPs in varied dose and time. Chitosan nanoparticles in median lethal concentration were 15 and 5 μg/ml for SUDHL-4 cells at 24 and 48 h, respectively (Figure 6). A high cytotoxicity against human lymphoma cell line could exert in chitosan nanoparticles. It was proved that the cell line is susceptible to CH-NPs by an IC_{50} value of 5 μg/ml after 48 h treatment and signifying that CH-NPs perhaps an excellent choice for antitumor drugs. Thus, size of the nanoparticles was mostly responsible in antitumor activity and in vivo distribution. Higher accumulation of tumour sites and makes longer half-life in vivo due to small size of nanoparticles [40- 43].

DNA Fragmentation

The cultured SUDHL-4 cells treated (i.e., 6 or 24 h) with 100 μg/ml CH-NPs were used for DNA extraction and detected necrosis by agarose gel electrophoresis. The DNA derivative with necrotic degeneration [44] was found after 6 h of incubation whereas larger DNA fragmentation was detected after 24 h of treatment with chitosan nanoparticles (Figure 7).

A series of "DNA Ladders" was observed in fragmented DNA due to formation of oligonucleosomes in apoptosis [45,46]. The fragmented DNA appears spectrum of size permanently during necrosis in contrast [43]. Thus, present study indicated that CH-NPs treated with tumour cells are mainly responsible for necrosis in fragmented DNA continuously.

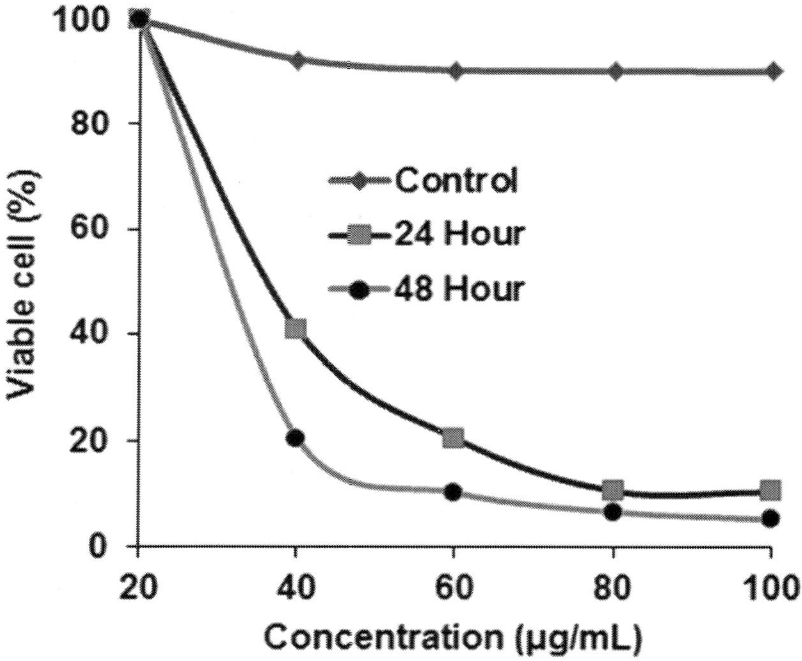

Figure 6: Inhibition of chitosan nanoparticles on SUDHL- 4 cell proliferation.

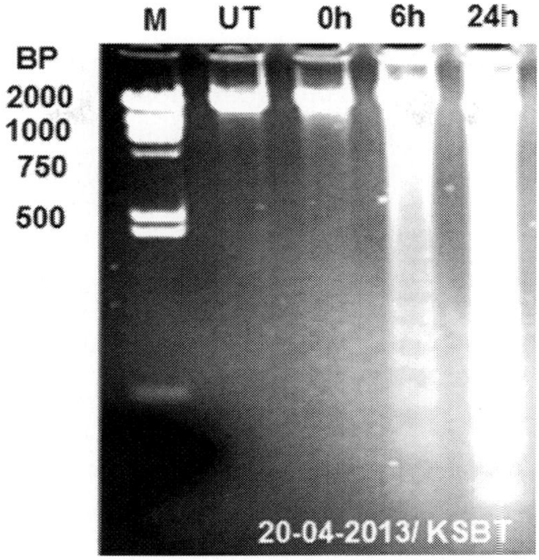

Figure 7: Showing agarose gel electrophoretic analysis of DNA isolated from SUDHL-4 cells incubated with 100 μg/mL chitosan nanoparticles for 6 h (lane 2) and 24 h (lane 3) or without treatment (lane 1). M: a DNA marker.

Alternations of Mitochondial Membrane Potential (MMP)

One of the promising mechanisms concerned for mitochondrial damage which is induced by CH-NPs in the necrosis of SUDHL-4 cells. In Present study, decrease mitochondrial membrane potential in SUDHL-4 cells was observed in different concentration of chitosan nanoparticles after 4 h of treatments. It was also observed that increased percentage of cells with loss of mitochondrial membrane potential extensively with more concentrations of CH-NPs. It was 74% with 100 μg/ml concentration of CH-NPs. The possible disruption of cell mitochondrial membrane and also strong disruption in mitochondrial membrane potential was observed after the treatment of chitosan nanoparticles (Figures 8(a) and (b)).

Regulation of cell death mainly takes place by mitochondria which were already reported [47]. In high Ca^{2+} and oxidative stress, mitochondria forced to changes which were swelling, de-energization, decrease mitochondrial potential and inner membrane permeabilization

[48]. In addition, necrotic cell death and loss of integrity in plasma membrane were observed due to dysfunction of mitochondria [49]. In this investigation, CH-NPs treated with lymphoma cells showed extreme decline of mitochondria membrane potential which indicated that the mitochondrial membrane is damaged (Figure 8(c)).

Effects of Cell Cycle

Different effects of CH-NPs treated with SUDHL-4 cells on cell cycle were determined by flow cytometry. The effects of CH-NPs were identified and a comparison was made in both treated and untreated cells on cell cycle. This finding was confirmed in the G_0/G_1 phase of cell cycle and also observed significant decrease of cells. The CH-NPs treatment with dose dependent manner showed apoptotic peaks and cell apoptotic incidence (9.9% in 100 μg/ml) [50]. The apoptotic incidence was increase up to 10.23% after treatment with 100 μg/ml chitosan nanoparticles (CH-NPs) (Figures 9(a) and (b)).This analysis exhibited that different concentrations of CH-NPs has major role for cell death and significant decline of cells in G_0/G_1 phase. This study also stated that CH-NPs showed high cytotoxicity towards lymphoma SUDHL-4 with necrotic cell death in about 52 mV positive surface charge and about 78 nm particle size. The CH-NPs and its antitumor mechanism are correlated with membrane disorderly inducing apoptosis activities.

CONCLUSION

In this investigation, we have developed a new method for synthesis of CH-NPs using ionic gelation technique

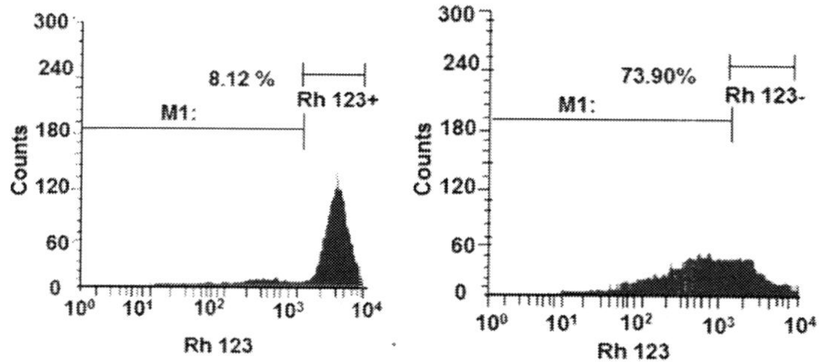

(a)

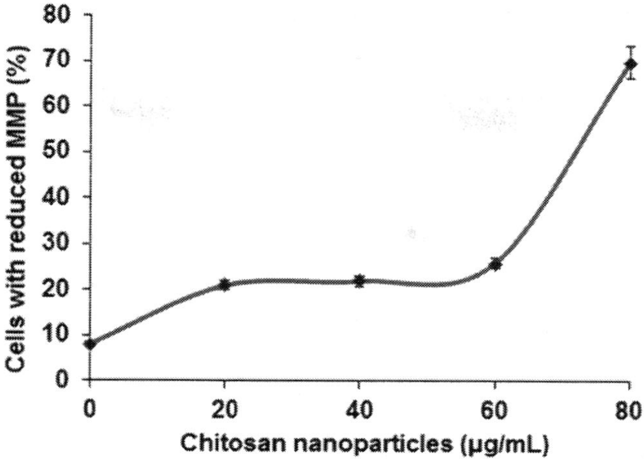

(b)

Figure 8: (a) Histogram of untreated cells; (b) Histogram of cells treated with 100 µg/ml chitosan nanoparticles; (c) Chitosan nanoparticles-induced changes of MMP (loss of MMP).

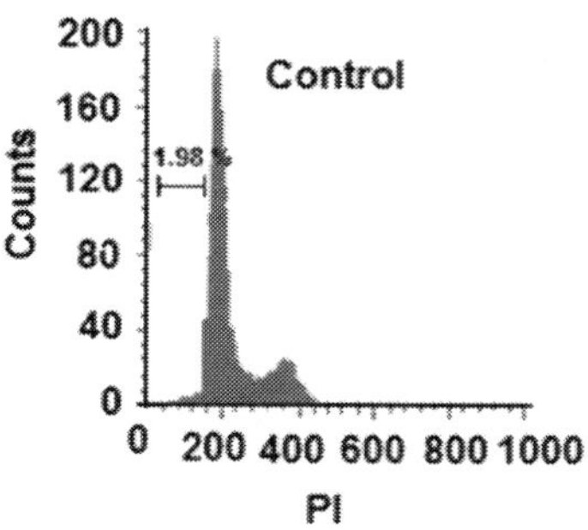

(a)

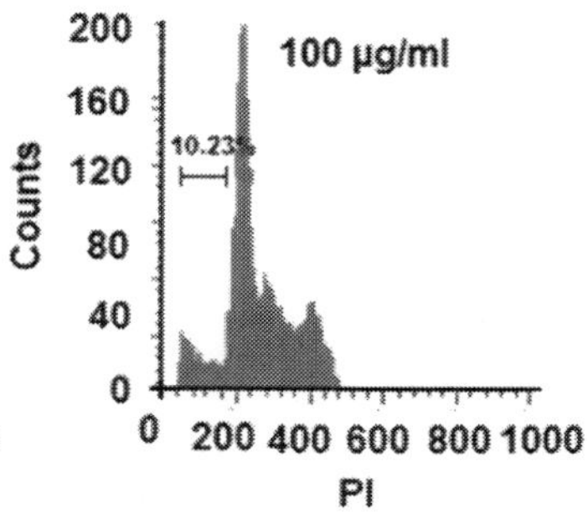

(b)

Figure 9: (a) Effect of chitosan nanoparticles on cell cycle of SUDHL-4 cells; (b) Apoptotic incidence of SUDHL-4 cells.

with TPP. The high loading competence with an efficient carrier was observed in the optimization ratio of chitosan to TPP (4:1). The present study demonstrated that a high cytotoxicity against human lymphoma SUDHL-4 cell line could exert in chitosan nanoparticles. It was proved that the cell line is susceptible to CH-NPs by an IC_{50} value of 5 µg/ml after 48 h action signifying that CHNPs are perhaps an excellent choice for antitumor drugs. Thus, size of the nanoparticles was mostly responsible in antitumor activity and in vivo distribution. Higher accumulation of tumour sites and makes longer half-life in vivo due to small size of nanoparticles. Here, the antitumor activity which prolongs efficiency could be controlled by using 78 nm particle size of CH-NPs. The CHNPs showed both lower as well as higher cationic charge densities which have higher cytotoxicity effects. The major factor for this cytotoxicity effect of chitosan derivatives is because of electrostatic ionic interface among the groups of negatively charged tumour cells and positively charged amino groups of chitosan. Therefore, the CH-NPs with 52 mV surface charge are responsible to make it higher cytotoxicity activity. Furthermore, the CH-NPs inhibited effectively the proliferation of Lymphoma SUDHL-4 in vitro through multiple mechanisms. These novel findings strengthen our hypothesis and can open a new path for further research against human lymphoma.

ACKNOWLEDGEMENTS

Umesh K. Parida, Niranjan Rout, Birendra K. Bindhani Authors are thanks to CSIR, Govt. of India for providing financial support. Authors are also grateful to CIPET, Bhubaneswar and Chennai of India for their help in this research. Authors have no any conflicts of interest in this manuscript. No writing assistance was utilized for preparation of this manuscript.

REFERENCES

1. Parida, U.K., Nayak, A.K., Bindhani, B.K. and Nayak, P.L. (2011) Synthesis and characterization of chitosan-polyvinyl alcohol blended with cloisite 30B for controlled release of the anticancer drug curcumin. Journal of Biomaterials and Nanobiotechnology, 2, 414-425. http://dx.doi.org/10.4236/jbnb.2011.24051
2. Qian, C., Xu, X., Shen, Y., Li, Y. and Guo, S. (2013) Synthesis and preliminary cellular evaluation of phosphonium chitosan derivatives as novel non viral vector. Carbohydrate Polymers, 97, 676-683.
3. Shukla, S.K., Mishra, A.K., Arotiba, O.A. and Mamba, B.B. (2013) Chitosan-based nanomaterials: A state-ofthe-art review.

International Journal of Biological Macromolecules, 59, 46-58.

4. Di Martino, A., Sittinger, M. and Risbud, M.V. (2005) Chitosan: A versatile biopolymer for orthopaedic tissueengineering. Biomaterials, 26, 5983-5990.http://dx.doi.org/10.1016/j.biomaterials.2005.03.016
5. Senel, S. and Mc Clure, S.J. (2004) Potential applications of chitosan in veterinary medicine. Advanced Drug Delivery Reviews, 56, 1467-1480.http://dx.doi.org/10.1016/j.addr.2004.02.007
6. Khor, E. and Lim, L.Y. (2003) Implantable applications of chitin and chitosan. Biomaterials, 24, 2339-2349. http://dx.doi.org/10.1016/S0142-9612(03)00026-7
7. Bumgardner, J.D., Wiser, R., Gerard, P.D., Bergin, P., Chestnutt, B., Marini, M., Ramsey, V., Elder, S.H. and Gilbert, J.A. (2003) Chitosan: Potential use as a bioactive coating for othopaedic and craniofacial/dental implants. Journal of Biomaterials Science, Polymer Edition, 14, 429-438. http://dx.doi.org/10.1163/156856203766652048
8. Kumar, M.N.V.R. (2000) A review of chitin and chitosan applications. Reactive and Functional Polymers, 46, 1- 27. http://dx.doi.org/10.1016/S1381-5148(00)00038-9
9. Synowiecki, J. and Khateeb, N.A. (2003) Production, properties and some new applications of chitin and its derivatives. Critical Reviews in Food Science and Nutrition, 43, 145-171. http://dx.doi.org/10.1080/10408690390826473
10. Shon, D.H. (2001) Chitosan oligosaccharides for functional foods and microbial enrichment of chitosan oligosaccharides in soy-paste. Proceedings of the International Workshop on Bioactive Natural Products, Tokyo, 56-66.
11. Tharanathan, R.N. and Kittur, F.S. (2003) Chitin—The undisputed biomolecule of great potential, Critical Reviews in Food Science and Nutrition, 43, 61-87.http://dx.doi.org/10.1080/10408690390826455
12. Sekiguchi, S., Miura,Y., Kancko, H., Nishimura, S.L., Nishi, N., Iwase, M. and Tokura, S. (1994) Molecular weight dependency of antimicrobial activity by chitosan oligomers. In: Nishimuri E. and Doi, E., Eds., Food Hydrocolloids: Structure, Properties and Function, Plenum Press, New York, 71-76.
13. Jeon, Y.J., Park, P.J. and Kim, S.K. (2001) Antimicrobial effect

of chitooligosaccharides produced by bioreactor. Carbohydrate Polymers, 44, 71-76.http://dx.doi.org/10.1016/S0144-8617(00)00200-9

14. Kondo, Y., Nakatani, A., Hayashi, K. and Ito, M. (2000) Low molecular weight chitosan prevents the progression of low dose streptozotocin induced slowly progressive diabetes mellitus in mice. Biological & Pharmaceutical Bulletin, 23, 1458-1464.http://dx.doi.org/10.1248/bpb.23.1458
15. Tsai, G.J., Wu, Z.Y. and Su, W.H. (2000) Antibacterial activity of a chitooligosaccharide mixture prepared by cellulase digestion of shrimp chitosan and its application in milk preservation. Journal of Food Protection, 63, 747-752.
16. Tsai, G.J., Wu, Z.Y. and Su, W.H. (2000) Antibacterial activity of a chitooligosaccharide mixture prepared by cellulase digestion of shrimp chitosan and its application in milk preservation. Journal of Food Protection, 63, 747-752.
17. Richardson, S.C.W., Kolbe, H.V.J. and Duncan, R. (1999) Potentials of low molecular mass chitosan as a DNA delivery system: Biocompatibility, body distribution and ability to complex and protect DNA. International Journal of Pharmaceutics, 178, 231-243.
18. Vimal, S., Abdul Majeed, S., Taju, G., Nambi, K.S., Sundar Raj, N., Madan, N., Farook, M.A., Rajkumar, T., Gopinath, D. and Sahul Hameed, A.S. (2013) Chitosan tripolyphosphate (CS/TPP) nanoparticles: Preparation, characterization and application for gene delivery in shrimp. Acta Tropica, 128, 486-493.
19. Oliveira, A.V., Silva, A.P., Bitoque, D.B., Silva, G.A. and Rosa da Costa, A.M. (2013) Transfection efficiency of chitosan and thiolated chitosan in retinal pigment epithelium cells: A comparative study. Journal of Pharmacy and Bioallied Sciences, 5, 111-118.
20. Qi, L.F., Xu, Z.R., Li, Y., Jiang, X. and Han, X.Y. (2005) In vitro effects of chitosan nanoparticles on proliferation of human gastric carcinoma cell line MGC803 cells. World Journal of Gastroenterology, 11, 5136-5141.
21. Singh, D.J., Lohade, A.A., Parmar, J.J., Hegde, D.D., Soni, P., Samad, A. and Menon, M.D. (2012) Development of Chitosan-based Dry Powder inhalation System of Cisplatin for Lung Cancer.

Indian Journal of Pharmaceutical Sciences, 74, 521-526.

22. Del Turco, S., Ciofani, G., Cappello. V., Gemmi, M., Cervelli, T., Saponaro, C., Nitti, S., Mazzolai, B., Basta, G. and Mattoli, V. (2013) Cytocompatibility evaluation of glycol-chitosan coated boron nitride nanotubes in human endothelial cells. Colloids Surf B: Biointerfaces, 111C, 142-149. http://dx.doi.org/10.1016/j.colsurfb.2013.05.031

23. Paiva, D., Ivanova, G., do Carmo Pereira, M. and Rocha, S. (2013) Chitosan conjugates for DNA delivery. Physical Chemistry Chemical Physics, 15, 11893-11899.

24. Liang, X., Li, X., Chang, J., Duan, Y. and Li, Z. (2013) Properties and evaluation of quaternized chitosan/lipid cation polymeric liposomes for cancer-targeted gene delivery. Langmuir, 29, 8683-8693. http://dx.doi.org/10.1021/la401166v

25. Sahu, A., Goswami, P. and Bora U. (2009) Microwave mediated rapid synthesis of chitosan. Journal of Materials Science: Materials in Medicine, 20, 171-175.http://dx.doi.org/10.1021/la401166v

26. Cheng, C.Y. and Li, Y.K. (2000) An Aspergillus chitinase with potential for largescale preparation of chitosan oligosaccharides, Biotechnol. Biotechnology and Applied Biochemistry, 32, 197-203. http://dx.doi.org/10.1042/BA20000063

27. Baxter, A., Dillon, M. and Anthony Taylor, K.D. (1992) Improved method for i.r. determination of the degree of N-acetylation of chitosan. International Journal of Biological Macromolecules, 14, 66-69.

28. Lavertu, M., Xia, Z., Serrequ, A.N., Berrada, M., Rodrigues, A., Wang, D., Buschmann, M.D. and Gupta, A. (2003) A validated 1H-NMR method for the determination of the degree of deacetylation of chitosan. Journal of Pharmaceutical and Biomedical Analysis, 32, 1149-1158.

29. Venkatesan, P., Puvvada, N., Dash, R., Prashanth Kumar, B.N., Sarkar, D., Azab, B., Pathak, A., Kundu, S.C., Fisher, P.B. and Mandal, M. (2011) The potential of celecoxib-loaded hydroxyapatite-chitosan nanocomposite for the treatment of colon cancer. Biomaterials, 32, 3794- 3806. http://dx.doi.org/10.1016/j.biomaterials.2011.01.027

30. Huang, Y. and Lapitsky, Y. (2011) Monovalent salt enhances

colloidal stability during the formation of chotosan/tripolyphosphate microgels. Langmuir, 27, 10392- 10399.

31. Jeevitha, D. and Amarnath, K. (2013) Chitosan/PLA nanoparticles as a novel carrier for the delivery of anthraquinone: Synthesis, characterization and in vitro cytotoxicity evaluation. Colloids Surfaces B: Biointerfaces, 101, 126-134.
32. Kim, H., You, S., Kong, B.W., Foster, L.K., Farris, J. and Foster, D.N. (2001) Necrotic cell death by hydrogen peroxide in immortal DF-1 chicken embryo fibroblast cells expressing deregulated MnSOD and catalase. Biochimica et Biophysica Acta, 1540, 137-146.http://dx.doi.org/10.1016/S0167-4889(01)00131-8
33. Behl, G., Sharma, M., Dahiya, S., Chhikara, A. and Chopra, M. (2011) Synthesis, characterization, and evaluation of radical scavenging ability of ellagic acid-loaded nanogels. Journal of Nanomaterials, 1-9. http://dx.doi.org/10.1155/2011/695138
34. Zhang, Z.P. and Feng, S.-S. (2006) The drug encapsulation efficiency, in vitro drug release, cellular uptake and cytotoxicity of paclitaxel-loaded poly (lactide)-tocopheryl polyethylene glycol succinate nanoparticles. Biomaterials, 27, 4025-4033.http://dx.doi.org/10.1016/j.biomaterials.2006.03.006
35. Stacey, N.B., Samantha, M.Y., Karl, R.F., Areti, T., Omid, K. and Ipsita, A.B. (2011) Ellagic acid promoted biomimetic synthesis of shape-controlled silver nanochains. Nanotechnology, 22, 1-10. http://dx.doi.org/10.1088/0957-4484/22/22/225605
36. Sonaje, K., Italia, J.L., Sharma, G., Bhardwaj, V., Tikoo, K. and Ravi Kumar, M.N.V. (2007) Development of biodegradable nanoparticles for oral delivery of ellagic acid and evaluation of their antioxidant efficacy against cyclosporine A-induced nephrotoxicity in rats. Pharmaceutical Research, 24, 899-908. http://dx.doi.org/10.1007/s11095-006-9207-y.
37. Gautam, B., Monal, S., Saurabh, D., Aruna, C. and Madhu, C. (2011) Synthesis, characterization, and evaluation of radical scavenging ability of ellagic acid-loaded nanogels. Journal of Nanomaterials, 2011, Article ID: 695138.http://dx.doi.org/10.1155/2011/695138
38. Mostafa, A.D., Adel Zaki, E., Mohamed, M.A. and Dina, M.D.B. (2012) Thermal stability and degradation of chitosan modified by

cinnamic acid. Open Journal of Polymer Chemistry, 2, 14-20. http://dx.doi.org/10.4236/ojpchem.2012.21003

39. Zobir, M.H., Samer, H.A.A., Zulkarnain, Z. and Muhammad, N.H. (2011) Development of antiproliferative nanohybrid compound with controlled release property using ellagic acid as the active agent. International Journal of Nanomedicine, 6, 1373-1383.
40. Yokoyama, M., Satoh, A., Sakurai, Y., Okano, T., Matsumura, Y., Kakizoe, T. and Kataoka, K. (1998) Incorporation of water-insoluble anticancer drug into polymeric micelles and control of their particle size. Journal of Controlled Release, 55, 219-229.
41. Zhang, L.Y., Hu, Y., Jiang, X.Q., Yang, C.Z., Lu, W. and Yang, Y.H. (2004) Camptothecin derivative-loaded poly (caprolactone-co-lactide)-b-PEG-b-poly(caprolactone-colactide) nanoparticles and their biodistribution in mice. Journal of Controlled Release, 96, 135-148. http://dx.doi.org/10.1016/j.jconrel.2004.01.010
42. Jinno, H., Ikeda, T., Matsui, A., Kitagawa, Y., Kitajima, M., Fujii, H., Nakamura, K. and Kubo, A. (2002) Section 5. Breast Sentinel lymph node biopsy in breast cancer using technetium-99m tin colloids of different sizes. Biomedicine and Pharmacotherapy, 56, 213-216.
43. Takenaga, M. (1996) Application of lipid microspheres for the treatment of cancer. Advanced Drug Delivery Reviews, 20, 209-219.
44. Yuan, X., Yang, X., Cai, D., Mao, D., Wu, J., Zong, L. and Liu, J. (2008) Intranasal immunization with chitosan/pCETP nanoparticles inhibits atherosclerosis in a rabbit model of atherosclerosis. Vaccine, 26, 3727-3734.http://dx.doi.org/10.1016/j.vaccine.2008.04.065
45. Morrison, M.R. and Smith, M. (1963) Preparation of fatty acid methyl esters and dimethyl acetyls from lipids with boron fluoride methanol. Journal of Lipid Research, 5, 600- 608.
46. Focher, B., Naggi, A., Torri, G., Cosani, A. and Terbojevich, M. (1992) Chitosans from Euphausia superba. 2: Characterization of solid-state structure. Carbohydrate Polymers, 18, 43-49.
47. Monti, M.G., Ghiaroni, S., Marverti, G., Montanari, M. and Moruzzi, M.S. (2004) Polyamine depletion switches the form of 2-deoxy-d-ribose induced cell death from apoptosis to necrosis in HL-60 cells.

The International Journal of Biochemistry & Cell Biology, 36, 1238-1248.

48. Vishu Kumar, A.B., Varadaraj, M.C., Lalitha, R.G. and Tharanathan, R.N. (2005) Characterization of chito-oligosaccharides prepared by chitosanolysis with the aid of papain and pronase, and their bactericidal action against Bacillus cereus and Escherichia coli. Biochemical Journal, 391, 167-175. http://dx.doi.org/10.1042/BJ20050093

49. Kanjanathaworn, N., Polpanich, D., Jangpatarapongsa, K. and Tangboriboonrat, P. (2013) Reduction of cytotoxicity of natural rubber latex film by coating with PMMA-chitosan nanoparticles. Carbohydrate Polymers, 97, 52-58.http://dx.doi.org/10.1016/j.carbpol.2012.12.078

50. Yang, S.J., Chang, S.M., Tsai, K.C., Tsai, H.M., Chen, W.S. and Shieh, M.J. (2012) Enhancement of chitosan nanoparticle-facilitated gene transfection by ultrasound both in Vitro and in Vivo. Journal of Biomedical Materials Research. Part B, Applied Biomaterials, 100, 1746-1754. http://dx.doi.org/10.1002/jbm.b.32741

14

CHAPTER

PREPARATION AND CHARACTERIZATION OF NOVEL DRUG-INSERTED-MONTMORILLONITE CHITOSAN CARRIERS FOR OCULAR DRUG DELIVERY

Dongzhi Hou[1], Ruyi Gui[1], Sheng Hu[1], Yi Huang[1], Zuyong Feng[2], Qineng Ping[3]

[1]College of Pharmacy, Guangdong Pharmaceutical University, Guangzhou, China

[2]School of Physics and Optoelectronic Engineering, Guangdong University of Technology, Guangzhou, China

[3]College of Pharmacy, China Pharmaceutical University, Nanjing, China

ABSTRACT

The objective of the study was to investigate the potential of montmorillonite as a sustained carrier in the preparation of drug-loaded nanoparticles for prolonged ocular application. Nanoparticles were prepared by ionic gelation of chitosan with sodium tripolyphosphate

(TPP). Betaxolol hydrochloride (BH) was applied as a model drug. TG, FT-IR, TEM, DLS and XRD have been employed in the characterization of drug-encapsulated chitosan-montmorillonite/TPP nanoparticles (CS-MMT/TPP NPs). TEM images showed that the particles were spherical in shape and had a rough surface. The size range of the nanoparticles was between 338 and 585 nm with positive zeta potential values from 24 mV to 36 mV and encapsulation efficiency values ranging from 12.27% to 50.92%. In vitro sustained drug release was observed with the BH-loaded nanoparticles in artificial tears (pH 7.4). The results of FT-IR, TG and XRD showed that the drug was coated with CS- MMT/TPP NPs. In the mucoadhesion studies, an interaction was found between drug-loaded CS- MMT/TPP NPs and mucin, which could enhance precorneal residence time and hence facilitate an effective sustained release. The optimized formulation was determined to be non-irritant and tolerable by modified Draize test. Therefore, the BH-loading CS-MMT/TPP NPs developed are a promising carrier for controlled drug delivery to the eye.

Keywords: Chitosan, Montmorillonite, Nanoparticles, Mucoadhesive, Sustained Release, Ocular Irritation

INTRODUCTION

It is worth mentioning that primary open-angle glaucoma is the most widespread neuropathy, affecting 60.5 million people in the world [1] . Aside from surgery, the only effective treatment for this disease is IOP-reducing drugs applied topically onto the cornea. The necessity of numerous daily applications of eye-drops for life is serious for patients, besides the risk of side effects in the anterior structures of the eye. Betaxolol hydrochloride (BH), and beta-adrenergic blocking agents, are indicated in the treatment of ocular hypertension and chronic open-angle glaucoma [2] . However, there are a number of challenges in the aspect of the treatment of ocular disease due to certain physiological process which may contribute to the poor efficacy of conventional drug formulations. For instance, blinking and tear drainage through the lachrymal drainage system can act to reduce the residence time of topically applied molecule. Eye-drops placed onto the ocular surface are washed in less than 30 seconds [3] . The corneal epithelium is almost impermeable to any substance larger than 500 Da [4] , and most of the commonly used topical drugs are larger than that and consequently cannot cross the cornea. Instead, they permeate throughout the conjunctiva and the underlying sclera in what is known as "non-productive passage". Altogether, less than 5% of topically administered drugs reach intraocular tissues [5] . Therefore, developing and designing a novel delivery platform, which suits the requirement of the topical ocular route is necessary.

Chitosan (CS), a cationic polysaccharide polymer, has some important properties such as mucoadhesivity, biocompatibility, nontoxicity and good ocular tolerance [6] . The mucoadhesive character of the polymer makes it increase the drug residence time due to the capacity to adhere to the mucin coat covering the conjunctiva and the corneal surfaces of the eye by non-covalent bonds forming the basis of ocular mucoadhesion [7] [8] . Moreover, CS solutions show pseudoplastic and viscoelastic properties [9] [10] , which are desirable for ocular drug administration. It has widely being used in ophthalmic preparations [11] . The report by R. S. Bhatta et al. showed that natamycin-loaded chitosan nanoparticles had the ability to prolong the retention time of drug on the ocular surface [12] . The efficiency of the pharmaceutical treatment of most ocular diseases is limited by short ocular residence of conventional pharmaceutical forms of drugs on the ocular mucosa due to the elimination of drug from the corneal surface by the lachrymal flow. To overcome the problem, ocular administration of drugs in the forms of CS nanoparticles has been proposed. CS-based systems can be used for improving the retention and distribution properties of drugs applied topically in the eye. Thus, all preferable properties make it a unique material for the design of ocular delivery vehicles [13] .

Indeed, taking into account the fact that cornea and conjuctiva have a negative charge, it was thought that the use of mucoadhesive polymers which might interact with these extraocular structures would increase the concentration and residence time of the associated drug [14] . The study on cyclosporin A-loaded chitosan nanoparticles has shown the interaction and prolonged residence time of chitosan nanoparticles at the ocular mucosa after the topical administration to rabbits [15] , and gatifloxacin-loaded SA-CH nano-reservoir system has also been investigated for the treatment of ocular infections [14] .

Montmorillonite (MMT) can form many nanocompsites with different organic compounds. Therefore, it expresses a significant capability to be used as a drug carrier for pharmaceutical purpose [16] . The report by Lin et al. has shown that 5-fluorouracil (5-FU) was successfully intercalated into the interlayer of montmorillonite [17] . In addition, the pillared MMT can be intercalated with ibuprofen as a sustained release carrier [18] . Therefore, MMT material has been proposed as a carrier with high potential for ocular drug delivery.

According to the reports [19] [20] , ionic gelation is the most studied and widely used method for fabricating CS nanoparticles, in which cationic chitosan and multivalent polyanions interact to form CS nanoparticles under simple and mild conditions. Among various polyanions, TPP is the most investigated one due to its quick gelling capacity and non-toxicity property. In the present study, the ionic gelation method was used to prepare BH- loaded CS-MMT/TPP NPs. Different fabricating conditions were evaluated in terms of their effects on physicochemical properties of BH-loaded CS-MMT/TPP NPs, including morphology, particle size, surface charge, the encapsulation efficiency, loading capacity, as well as release profile. MMT and chitosan were chosen as sustained carriers to achieve better encapsulation efficiency and controlled release profiles of optimized nanoparticles. Therefore, the purpose of this study was to develop a suitable and mucoadhesive system consisting of MMT, chitosan and BH for ocular drug delivery.

EXPERIMENTAL SECTION

Materials

Chitosan (deacetylation degree of 95%) was purchased from HAIDEBEI MARINE BIOEEING CO. LTD, Jinan, China. Momtmorillonite was obtained from Zhejiang sanding CO. LTD, china. Betaxolol hydrochloride was purchased from Jinan Haohua Industry, Jinan , China . Male, New Zealand albino rabbits of 2.2 - 2.5 kg were provided by laboratory animal center, Traditional Chinese medicine university of Guangzhou. All other chemicals were of analytical grade.

Acid Activation of Momtmorillonite

The MMT was activated by treating with 5% H_2SO_4 at the temperature of 40°C for 1.5 h. The resulting clays were centrifuged and washed with water several times until they were free of SO_4^{2-}, and then were dried at 100 °C in an air oven until constant weight was attained.

Preparation of CS-MMT/TPP NPs

The nanoparticles were prepared by ionic gelation techniques. In the first step MMT was added to acetum (pH = 4.8, 2% v/v) containing betaxolol hydrochloride and heated in water bath at 60°C for 6 hours, and then the solution was ultrasonicated for 5 min and added to chitosan solution and stirred for 1.5 h. subsequently the nanoparticles were fabricated by adding above MMT suspension to chitosan solution (2% v/v) while the TPP solution was added to the mixture suspension with ultrasonication for 10 s and stirring for 60 mins. The dispersed nanoparticles were frozen at −80°C for 48 h, and then freeze-dried at −50°C for 12 h. Finally, the freeze-dried powders were stored in glass vials for further study.

Entrapment Efficiency (EE) and Drug Loading Capacity (DLC)

To determine the free BH during the preparation of BH-loading CS/TPP NPs and CS-MMT/TPP NPs, the dialyzed solution was collected. The concentration of free BH in the solution was determined by UV-VIS spectrophotometry (UV-1800, MAPADA instrument CO.LTD) using a wavelength of 274.5 nm. The amount of BH in nanoparticles was calculated as the difference between the total amount used to prepare the nanoparticles and the amount present in the supernatant. Fresh prepared BH-loaded CS-MMT/TPP NPs were used for determining the drug-loading capacity. NPs were dissolved in HPLC grade methanol by stirring for 4h, and the solution was ultrasonicated for 5 mins. This colloidal dispersion was filtered by 0.22 μm membrane filter. The drug concentration was analyzed by UV-VIS. The EE and LC of nanoparticles were determined in triplicate and calculated as follows:

$$\text{Encapsulation Efficiency}\left(\text{EE\%}\right)=\frac{\left(\text{Total amount of BH-Free BH}\right)}{\text{Total amount BH}}\times 100\% \tag{1}$$

$$\text{Drug loading capacity}\left(\text{DLC\%}\right)=\frac{\left(\text{Total amount of BH-Free BH}\right)}{\text{Total weight NPs}}\times 100\% \qquad (2)$$

In Vitro Release Studies

The release of BH from nanoparticles was studied by dialysis method in pH 7.4 artificial tears. 2 ml samples were instilled in the dialysis bag which was screwed with two clamps at each end. The dialysis bag was dipped into the receptor compartment containing 35 ml of dissolution medium and stirred continuously at 100 rpm and maintained at 34°C. The receptor compartment was closed to prevent evaporation of the dissolution medium. Samples were withdrawn at regular time intervals, and the same volume was replaced with fresh dissolution medium. The samples were spectrophotometricially measured at 274.5 nm.

Particle Size and Zeta Potential

NPs size distribution and zeta potential were determined using photon correlation spectroscopy with a Zetasizer 3000 (DelsaNano S, USA). The size distribution analysis was performed at a scattering angle of 90° at a temperature of 25°C, and zeta potential was measured by electrophoretic light scattering using a disposable zeta cuvette. All measurements were performed in triplicate and the results were presented as mean ± SD.

Nanoparticles Morphonology

The morphological examination of nanoparticles was performed on transmission electron microscopy (TEM) (JEM-2010HR, Japan). The nanoparticles were applied to a carbon-coated 300 copper grid and then were stained with 2% phosphotungstic acid for viewing by TEM.

Fourier Transform Infrared Spectroscopy (FT-IR)

FT-IR spectra of freeze-dried nanoparticles were obtained with Bruker Vertex-70 spectrophotometer using potassium bromide (KBr) disk technique. FT-IR spectra of BH-loaded CS-MMT/TPP NPs and their pure components were compared for the evaluation of drug-excipient and excipient-excipient chemical interaction.

Thermogravimetric Analysis (TG)

TG analysis for BH, CS, MMT, acid-MMT, physical mixture and BH-loaded CS-MMT/TPP NPs were performed using a Netzsch STA 409PC TG instrument with NETZSCH-TA4_5 software Proteus Analysis (Germany). The apparatus was calibrated with purified indium (99%). The samples were placed in flat-bottomed aluminum pan and heated at a constant rate of 10°C/min in nitrogen atmosphere in a temperature range of 30°C - 1000°C.

X-Ray Diffratometry

X-ray diffraction patterns for BH, CS, acid-MMT, physical mixture, BH-loaded CS-MMT/TPP NPs were obtained using a Bruker D8 Advance X-ray diffractometer equipped with CuKα. The tube was operated at 40 KV and 40 mA.

Mucoadhesion Studies

Equal volumes of mucin solution and nanoparticles were vortexed for 1 min and zeta potential of the mixture was measured by photon correlation Spectroscopy with a Zetasizer 3000 (DelsaNano S, USA). In this study, the mucoadhesive properties of nanoparticles were evaluated by measuring the change of zeta potential on interaction with negatively charge mucin [21] . The alteration of zeta potential of nanoparticles indicates the interaction with mucin.

Ocular Irritation Studies

The ocular irritancy and damaging effects of NPs and BH were evaluated according to a modified Draize test [22] . Test substance (50 ul) was instilled directly to the cornea in the right eye every 12 h for 7 days (14 treatments), while left eye severed as control and was treated with normal saline. Observation was carried out before the treatment for every day and was done to evaluate ocular tissue after the last treatment of 1, 2, 4, 24, 48 and 72 h. The congestion, swelling, discharge and redness of the conjunctiva were grade on a scale from 0 to 3, 0 to 4, 0 to 3, and 0 to 3, respectively. Irritation and cornea opacity were grade on a scale from 0 to 4.

Statistical Analysis

The average difference of data was compared by one-way analysis of variance or independent sample t-test (origin 8.5 USA). All data are presented as a mean value with its standard deviation indicated (mean ± SD). P-values less than 0.05 were considered to be statistical significance.

RESULT AND DISCUSSION

Entrapment Efficiency (EE) and Drug Loading Capacity (DLC)

The effect of CS concentration on the encapsulation efficiency and drug-loading capacity were examined as shown in Table 1. The increasing concentration of CS from 1.0 mg/ml to 1.5 mg/ml at a constant TPP concentration and chitosan to TPP mass ratio, caused a great increase in drug content from 30.32% to 39.51%. This indicates that a less amount of nanoparticles were formed due to the lower chitosan concentration. However, when the concentration of CS was increased from 1.5 mg/ml to 3.0 mg/ml, a significant decrease in the encapsulation efficiency from 39.51% to 31.6% was exhibited. This may be explained by the fact that relatively higher viscosity of the higher CS concentration hinders the drug encapsulation [23] . With the similar results, the loading capacity firstly increased with the increase of the CS concentration and then decreased.

Table 1: Effect of chitosan concentration on the encapsulation efficiency and drug loading capacity for the prepared nanoparticles (mean ± SD)

Formula code	CS concentration (mg/ml)	Encapsulation efficiency (%)	Loading capacity (%)
1	1	30.32 ± 0.12	11.49 ± 0.33
2	1.5	39.51 ± 0.23	11.99 ± 0.26
3	2	33.32 ± 0.13	8.67 ± 0.45
4	2.5	33.01 ± 0.21	6.68 ± 0.78
5	3	31.16 ± 0.11	5.99 ± 0.13

Particles preparation conditions: the weight of montmorillonite = 5 mg, BH concentration = 1.25 mg/ml, chitosan/TPP = 10:3, pH = 4.8.

The reason appears to be similar to the above definition. In addition, the increase of the weight of CS and the corresponding change of encapsulation efficiency were a contributor to the result as well. The effects of chitosan to TPP mass ratio on the drug content and loading capacity of nanoparticles were studied at the mass ratio of 10:1, 10:1.5, 10:2, 10:2.5 and 10:3 at a fixed TPP concentration of 1.0 mg/ml, and the results were presented in Table 2. The encapsulation efficiency and loading capacity of nanoparticles decreased with the decrease in weight ratio of CS/TPP from 10:1 to 10:2. A possible explanation may be that a more compact solid-matrix structure due to the increased mass ratio for CS/TPP, which leads to the increasing amount of formed nanoparticles, resulting in the increase for encapsulation efficiency and loading capacity in nanoparticles. However, when the mass ratio is over the critical point (10:2), the aggregation of nanoparticles occurs, which indicated the reduction of encapsulation efficiency. The influence of the different amount of MMT on drug content and drug loading capacity was also examined, as shown in Table 3. Table 3 shows that encapsulation efficiency increased from 5.5% to 9.21%, when initial weight of MMT increased from 1 mg to 5 mg. However, the further increase of MMT amount does not lead to a sequential increase of encapsulation efficiency. The reason may be the more drugs adsorbed and intercalated with the higher MMT weight could be a contributor to the increase. When the amount of MMT was increased to some value, the drug content appears to keep a constant value possibly due to the redundant MMT which does not participate in the formation of the nanoparticles. The effect of initial BH concentration on drug content and loading capacity of the nanoparticles were also examined, as shown in Table 4. The increase of BH concentration was done from 0.75 mg/ml to 1.25 mg/ml, accompanied with the increasing encapsulation from 3.12% to 12.44% (Table 4). However, when BH concentration reaches 1.25 mg/ml, the drug content tends to decrease with progressive enhancement of drug loading capacity. This indicates that BH adsorption and encapsulation by electrostatic interactions normally exhibit a saturation kinetics which reaches a peak value at high concentration under constant temperature. On the basis of single-factor test, orthogonal test was carried out by choosing four factors and each factor had three levels (Table 5). Preparation technology was optimized through the orthogonal experiment. According to the analysis of variance, the influence factors of drug content of the nanoparticles were followed by the sequence: TPP concentration > chitosan concentration > drug concentration > chitosan to TPP mass ratio. The optimized preparation was chosen for further study.

Effect of Chitosan Concentration on In-Vitro Release Performance

Figure 1 shows BH release profiles in the respect of different chitosan concentration during BH encapsulation. The reduction of initial burst release with the chitosan concentration increasing from 1.0 mg/ml to 2.0 mg/ml during BH encapsulation was observed. This may be explained by the fact that the increased viscosity at higher chitosan concentration allowed the formation of denser nanoparticles upon the interaction with TPP, resulting in greater crosslink density and less swelling ability, and consequently reducing the BH release [24] . In addition, the decrease of encapsulation efficiency with the increase of chitosan concentration may be a contributor to this phenomenon. With the durative increase chitosan concentration, the increase of initial burst release for nanoparticles was observed due to the aggregation of nanoparticles caused by the increased chitosan concentration resulting in the more loose structure for nanoparticles.

Effect of Amount Varies of Montmorillonite on In-Vitro Release Performance

The effect of different amount of montmorillonite on BH release from prepared nanoparticles was studied. From

Table 2: Effect of chitosan to TPP mass ratio on the encapsulation efficiency and drug loading capacity for the prepared nanoparticles (mean ± SD)

Formula code	Mass ratio of CS to TPP	Encapsulation efficiency (%)	Loading capacity (%)
1	10:1	45.63 ± 0.11	13.87 ± 0.78
2	20:3	43.77 ± 0.36	12.64 ± 0.34
3	10:2	39..82 ± 0.23	11.58 ± 0.46
4	20:5	39.41 ± 0.21	11.49 ± 1.26
5	10:3	36.51 ± 0.15	11.15 ± 0.27

Particles preparation conditions: the weight of montmorillonite = 5 mg, BH concentration = 1.25 mg/ml, chitosan concentration = 2 mg/ml, pH = 4.8.

Table 3: Effect of the amount of MMT on the encapsulation efficiency and drug loading capacity for the prepared nanoparticles (mean ± SD)

Formula code	MMT (mg)	Encapsulation efficiency (%)	Loading capacity (%)
1	1	19.29 ± 0.25	5.5 ± 0.21
2	3	26.96 ± 0.32	7.55 ± 0.34
3	5	35.47 ± 0.12	9.21 ± 0.46
4	7	37.13 ± 0.45	9.38 ± 0.67
5	10	37.75 ± 0.79	9.48 ± 0.43

Particles preparation condition: chitosan concentration = 2 mg/ml, BH concentration = 1.25 mg/ml, chitosan/TPP = 10:3, pH = 4.8.

Table 4: Effect of various initial drug concentration on the encapsulation efficiency and drug loading capacity for prepared nanoparticles (mean ± SD)

Formula code	Drug concentration (mg/ml)	Encapsulation efficiency (%)	Loading capacity (%)
1	0.75	11.92% ± 0.13%	3.12% ± 0.12%
2	1.0	17.20% ± 0.21%	5.38% ± 0.25%
3	1.25	34.72% ± 0.11%	12.44% ± 0.01%
4	1.5	23.79% ± 0.26%	10.20% ± 0.77%
5	2.0	21.65% ± 0.33%	10.09% ± 0.39%

Particles preparation conditions: the weight of montmorillonite = 5 mg, chitosan concentration = 1.25 mg/ml, chitosan: TPP = 10:3. pH = 4.8.

Table 5: Effect of chitosan concentration, TPP concentration, mass ratio of chitosan/TPP and BH concentration on the encapsulation efficiency, loading capacity, particle size, zeta potential values and PID (mean ± SD)

code	A[a]	B[b]	C[c]	D[d]	EE (%)	LC (%)	Particle size (nm)	Zeta Potential (mV)	PID
1	1.0	0.8	10:1	0.75	13.10 ± 0.21	5.26 ± 0.16	338.4 ± 2.3	25.14 ± 3.6	0.254 ± 0.11
2	1.0	1.0	10:1.5	1.25	15.17 ± 0.11	8.20 ± 0.34	366.5 ± 5.6	30.31 ± 2.7	0.213 ± 0.08
3	1.0	1.2	10:2.5	2	18.00 ± 0.32	11.49 ± 0.31	585.5 ± 7.8	24.62 ± 1.2	0.282 ± 0.23
4	1.5	0.8	10:1.5	2	19.53 ± 0.23	10.56 ± 0.24	355.8 ± 4.3	29.49 ± 4.8	0.222 ± 0.34
5	1.5	1.0	10:2.5	0.75	12.27 ± 0.13	3.49 ± 0.38	529.3 ± 2.2	31.82 ± 2.1	0.260 ± 0.21
6	1.5	1.2	10:1	1.25	35.31 ± 0.42	14.85 ± 0.56	480.5 ± 4.5	30.70 ± 3.9	0.282 ± 0.13
7	2.0	0.8	10:2.5	1.25	36.11 ± 0.35	11.28 ± 0.43	503.5 ± 5.9	25.27 ± 4.2	0.242 ± 0.22
8	2.0	1.0	10:1	2	13.40 ± 0.12	6.33 ± 0.57	451.7 ± 6.7	29.40 ± 5.6	0.220 ± 0.26
9	2.0	1.2	10:1.5	0.75	50.92 ± 0.48	11.59 ± 0.24	524.3 ± 8.2	36.42 ± 2.3	0.275 ± 0.15

[a]Chitosan concentration; [b]TPP concentration; [c]Chitosan/TPP mass ratio; [d]BH concentration.

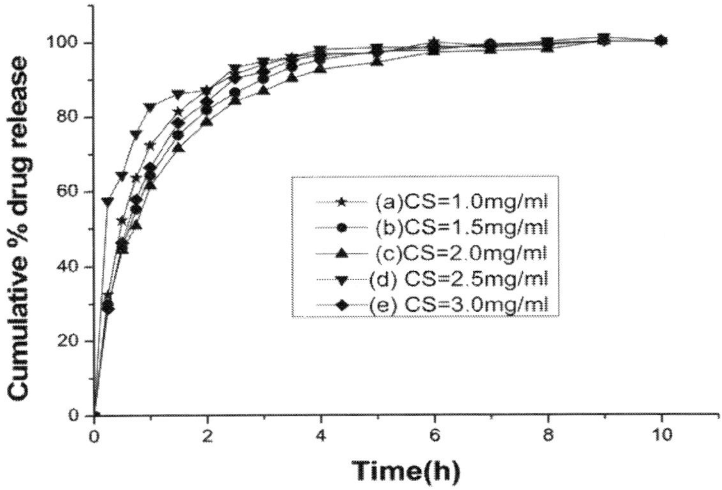

Figure 1: The influence of different chitosan concentration (a) chitosan = 1.0 mg/ml; (b) chitosan = 1.5 mg/ml; (c) chitosan = 2.0 mg/ml; (d) chitosan = 2.5 mg/ml; (e) chitosan = 3 mg/ml) on BH release from prepared nanoparticles.

the release profiles in Figure 2, we can see that the overall rate of drug release (i.e. the release fraction of drug content of nanoparticles at a given time) tended to become lower as the amount of MMT increasing from 1 mg to 5 mg, while it started to become higher with the further increase of the weight of MMT. It is obvious that the amount of MMT should be in an appropriate range, which can only effectively form nanoparticles with chitosan and TPP. This indicates that MMT at high amount would interrupter the formation of denser chitosan nanoparticles due to the hydroxyl group in the MMT interacting with the amino of chitosan, which disturbed the electrostatic interaction between chitosan and TPP. These reasons resulted in more fast release of BH from nanoparticles.

The result of drug release experiment for BH solution and optimized NPs in stimulated ocular circumstances (37°C, pH 7.4) are shown in Figure 3. We can observe that BH-loaded CS-MMT/TPP NPs showed a lower burst release as compared to BH-loaded CS/TPP NPs, but both presented a complete release of BH over a period of 10 h. For pure BH solution, which showed a fast release profile with 100% BH released within 2 h. Hence, NPs exhibited significantly sustained release in comparison to drug solution. In addition, this could also indicate that CS-MMT/TPP NPs showed better sustained-release compared to that of CS/TPP NPs. From the release profiles of NPs, we may come to a conclusion that BH encapsulated in the matrices of nanoparticles are released by three mechanisms taking place in sequence: 1) BH desorption from the particle surface; 2) diffusion and readsorption of the BH through the pores of the polymer network or the interlayer of montmorillonite; 3) degradation and erosion of the polymeric network.

Particle Size and Zeta Potential

The reduced size of nanoparticle formulation was very interesting in the view of their application in ocular drop formulations. It is well known that the size of particle under 10 μm would not cause any ocular irritation. Therefore, the reduction achieved in particles size could result in more compliance for patient. In this study, all the nanosystems ranged in diameter from 338.4 ± 2.3 to 585.5 ± 7.8 nm with a polydispersity index values between 0.213 ± 0.08 to 0.282 ± 0.13, as shown in Table 5.

The effect of concentration and weight ratio of CS and TPP on the mean nanoparticle diameter were examined. The nanoparticle diameter appears to be dependent on the concentration of CS and TPP and chitosan to TPP mass ratio, the minimum diameter (338.4 nm) corresponding to the lowest CS and TPP concentration, and maximum diameter (585.5 nm) corresponding to the highest TPP concentration and weight ratio between chitosan and TPP. The increase CS concentration from 1.0 mg/ml for formula 1 to 1.5 mg/ml for formula 4, then 2.0 mg/ml for formula 7 at the constant TPP concentration, showed a significant increase in the particle diameter from 338 to 355, then 503 nm, respectively. The increased particle diameters were attributed to the increase of

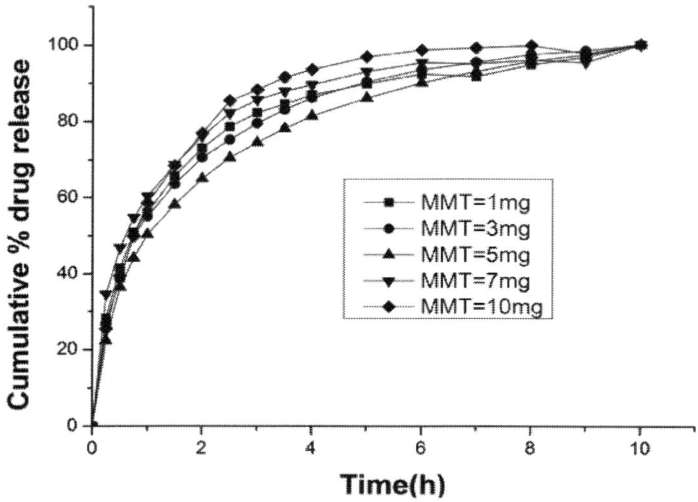

Figure 2: In vitro release profiles of different amount of acid MMT prepared nanoparticles.

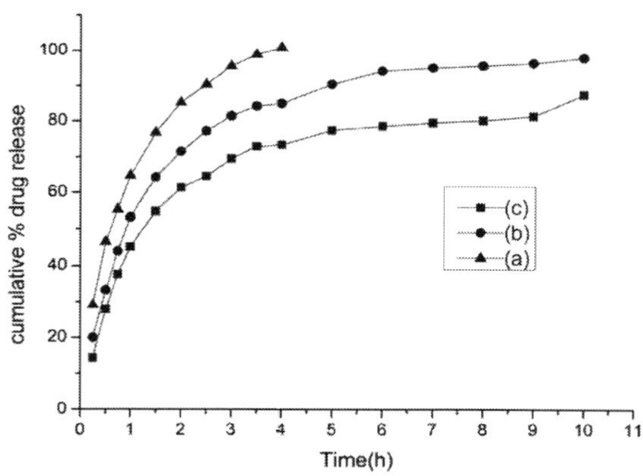

Figure 3: In vitro release profiles of (a) BH, (b) BH loading CS/TPP NPs,(c) BH loading CS-MMT/TPP NPs.

CS viscosity [25] and the higher availability of protonated amine groups for ionic gelation with increasing CS concentration [14] .

Zeta potential studies of the nanoparticles showed the values ranging from 24.62 ± 1.2 to 36.42 ± 2.3 mv (Table 5). The positive zeta-potential values indicate that the surface of the nanosystems is mostly composed of

CS [26] . This net positive charge of the particle is desirable to promote electrostatic interaction with the ocular mucin [27] . In general, it was noted that zeta potential of the nanoparticles decreased as the CS/TPP ratio decreased. The lower zeta potential with increasing TPP amounts might be caused by an increased masking of free positively changed amino groups of CS. In addition, increasing CS concentration results in an increase of zeta potential of the nanoparticles due to the increase of free positively charged amino groups of CS. According to the analysis of variance (data not shown), the effect factors on nanoparticles size and zeta potential take place in sequence: chitosan to TPP mass ratio > TPP concentration > chitosan concentration > drug concentration and chitosan to TPP mass ratio > chitosan concentration > TPP concentration > drug concentration, respectively.

Nanoparticles Morphology

A representative TEM image for drug-loaded CS-MMT/TPP nanoparticles was presented in Figure 4. The results indicated that the tested nanoparticles were predominantly spherical in the shape with an irregular surface. The diameter of nanoparticles was about 250 nm smaller than the diameter determined by photon correlation spectroscopy, in which this could have been expected since the fresh prepared nanoparticles suspensions for the photon correlation spectroscopy experiments, and the chitosan and MMT have the ability to swell in contact with acetic acid, while the TEM experiments were performed with dry samples.

Fourier Transforms Infrared Spectroscopy (FT-IR)

BH, chitosan, montmorillonite, their physical mixture, BH-loaded CS/TPP NPs and BH-loaded CS-MMT/TPP NPs were analyzed using FT-IR spectrophotometer for characteristic absorption bands (Figure 5). The IR

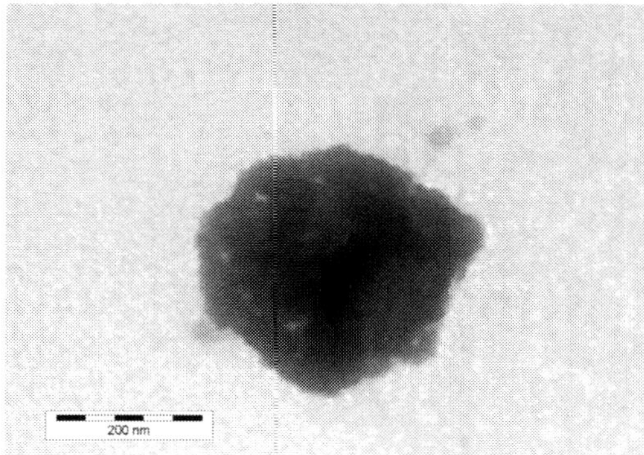

Figure 4: Transmission electron micrograph of BH-loaded CS-MMT/TPP NPs, bar represent 1 μm.

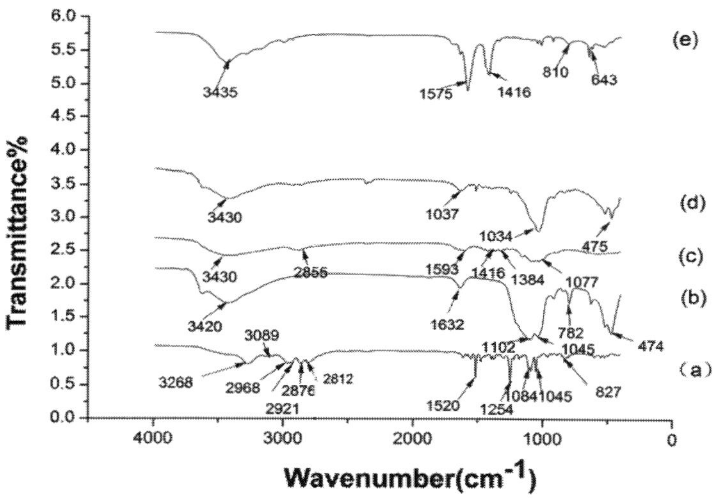

Figure 5: FTIR spectrum of (a) BH, (b) MMT, (c) CS, (d) Physical mixtures of BH, MMT and CS, (e) BH-loaded CS-MMT/TPP NPs.

spectrum of BH showed several obvious characteristic peaks: the strong peak at 3248 cm^{-1}belongs to the -OH stretching vibrations; aromatic C=C stretching vibrations are at 1610 cm^{-1} and 1520 cm^{-1}; aromatic C-H stretching vibrations are at 3089 cm^{-1}; the peaks at 2968 cm^{-1} and 2961 cm^{-1} correspond to $-CH_3$ stretching; the peaks at 2876 cm^{-1} and 2812 cm^{-1} correspond to $-CH_2$ stretching; C-O-C stretching vibrations are at

1084 cm^{-1}. In MMT spectrum the strong peak at 3420 cm^{-1} corresponds to the O-H stretching vibrations; Si-O stretching is at 1102 cm^{-1} and 1042 cm^{-1}; the intense peaks of 519 cm^{-1} and 479 cm^{-1} belong to Fe-O-Si stretching and Mg- O-Si stretching, respectively; the stretching of Mg-Al-OH is at 782 cm^{-1}; The IR spectrum of chitosan was characterized by the intense peak at 3435 cm^{-1} corresponded to the combined peak of O-H stretching and intermolecular hydrogen bonding. Specific peaks of chitosan were observed at 1630 cm^{-1} (amide I) and 1593 cm^{-1} (amine II). The symmetric stretch of C-O-C is found at around 1077 cm^{-1}, and the intense peak at 1416 cm^{-1} belongs to C-N stretching.

The spectrum patterns of physical mixtures corresponded simply to the superposition of the IR spectra of betaxolol, montmorillonite and chitosan. The characteristic peaks for BH were almost completely obscured by the very intense and broad peaks of chitosan and montmorillonite. The FT-IR spectrum of BH-loaded CS-MMT/ TPP NPs is different from that of physical mixtures and each pure component. In NPs the tip of peak at 3420 cm^{-1} was shifted to 3435 cm^{-1}, indicating that hydrogen bonding is enhanced. The 1953 cm^{-1} peak of N-H blending vibration disappeared and new peaks at 1575 cm^{-1} and 1416 cm^{-1} were observed. This spectra changed may be attributed to the electrostatic interaction between chitosan and TPP. In addition, the hydrogen-bond interaction between chitosan and montmorillonite could explain this fact as well.

Thermogravimetric Analysis (TG)

The TG curves of pure components, their corresponding physical mixture, BH-loaded CS/TPP NPs and BH- loaded CS-MMT/TPP NPs were showed in Figure 6. There are two distinct changes in the TG profiles of montmorillonite. The first weight loss from 42°C to 184°C may be due to the free water evaporation. The second weight loss from 480°C to 625°C is attributed to the structural dehydration. In the curves of simple physical mixture, the weight loss from 228°C to 347°C corresponds to the decomposition of BH. Other weight losses are the same as those of MMT and CS. The TG patterns of MMT-BH/CS NPs are shown in Figure 6(d). The first weight loss from 42°C to 184°C may be attributed to the free water evaporation. The second weight loss from 228°C to 347°C may be owed to free drug decomposition. The third weight loss from 450°C to 520°C is attributed to the drug decomposition which was encapsulated into nanoparticles. We can see that the weight loss at around 300°C in Figure 6(e) shifts to 500°C in Figure 6(d), which may be explained by the fact that the BH is partly encapsulated into the nanoparticles. The last weight loss from 840°C to 900°C may be due to the destruction of ionic bond between CS and TPP.

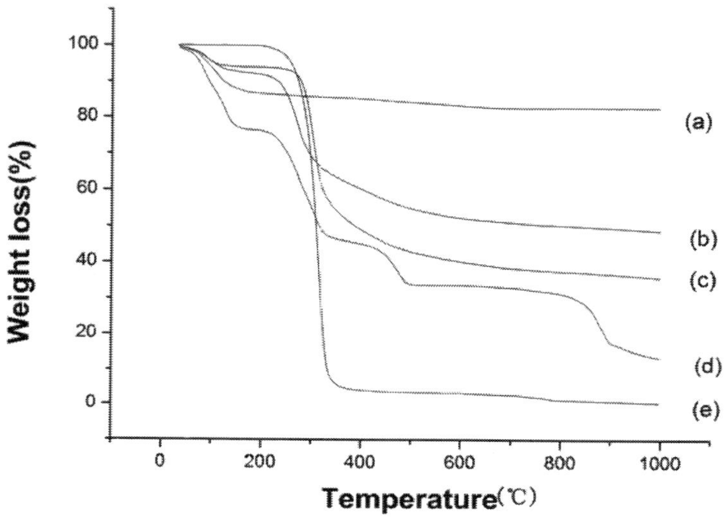

Figure 6: TGA curves of (a) MMT, (b) physical mixture of BH, MMT, CS, (c) CS, (d) BH loading CS-MMT/TPP NPs, (e) BH.

X-Ray Diffratometry

Figure 7 shows XRD patterns of BH, acid-MMT, CS, their physical mixture, blank nanoparticles and optimized nanoparticles. BH has a specific sharp peak. MMT and CS show the topical reflection at 2-theta = 5.82 and 20.18, respectively. No characteristic peaks of CS and MMT were found in diffracograms of blank nanoparticles and optimized nanoparticles, which indicated that chitosan nanoparticles were comprised of a dense network structure of interpenetrating polymer chain crosslinked to each other by the electrostatic reaction between TPP and chitosan. The XRD demonstrated the greater disarray in chain alignment in the nanoparticles after cross-link. Therefore, the BH-loaded CS-MMT/TPP NPs were successfully formed.

Mucoadhesion Studies

Mucoadhesion studies of the optimized nanoparticles showed that the nanoparticles revealed the reduction in zeta potential values for CS nanoparticles occurring on mixing with mucin (Figure 8). The reduction may be attributed to the ionic reaction between negatively charged sialic acid residues in mucin and positively charged

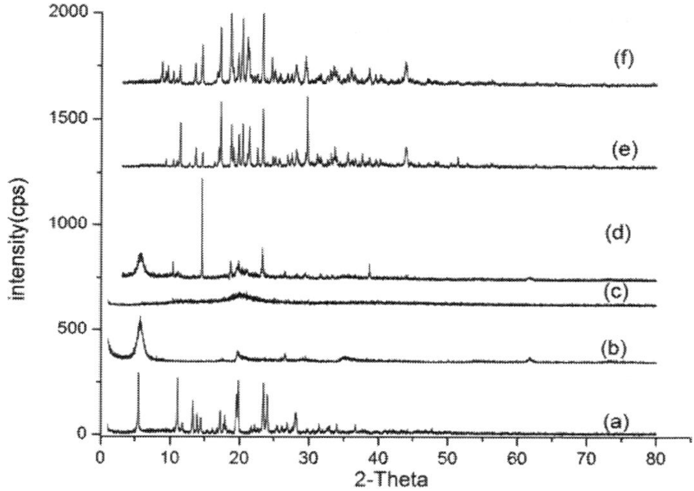

Figure 7: X-ray difffraction of (a) BH; (b) acid-MMT; (c) CS; (d) their physical mixture; (e) blank NPs and (f) oputimized NPs.

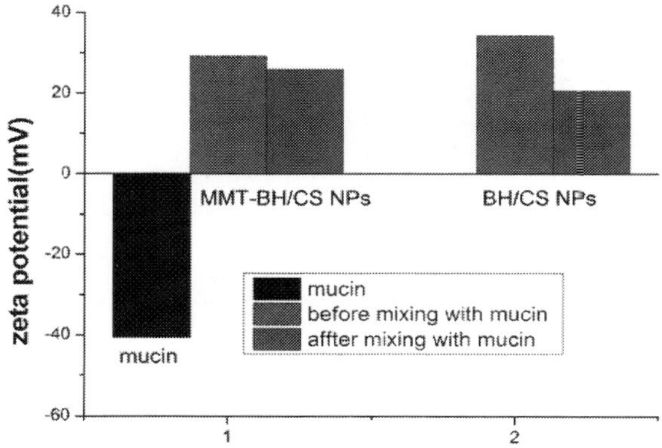

Figure 8: Zeta potential values obtained for mucin and BH/CS and MMT- BH/CS nanoparticles before and after incubation with mucin.

amino groups in CS nanoparticles that results in the mucoadhesive properties of the tested nanoparticles. BH/CS nanoparticles showed more reduction in zeta potential values indicating more mucoadhesive compared to BH- loaded CS-MMT/TPP NPs. This may be attributed to the fact that BH-loaded CS/TPP NPs had higher zeta potential values (+34 mV) than MMT-BH/CS nanoparticles (+29.18 mV).

Ocular Irritation Studies

Figure 9 presents the histopathology of the tested rabbit's eyeballs with various preparations to investigate their influence on cell structure and tissue integrity. The in vivo study showed no sign of irritation or damaging effects in cornea, conunctiva or iris. The scores for conjunctiva congestion of MMT-CS/TPP nanoparticle formulation and control group were grade zero. The cornea and iris scores were zero for all tested samples at all observations (data not shown). Pathology of rabbit eyeballs and lids confirmed the presence of normal ocular surface structure in both optimized nanoparticles cornea and conjunctival epithelial cell maintained normal morphology and integrated epithelium. The basal columnar cell of cornea can be recognized as single lager cell, and is normally packed by tight junction complex (Figure 9(a) & Figure 9(b)). Similarly, normal levels of scattered polymorphonuclear cells were observed in the conjuctival stroma (Figure 9(c) & Figure 9(d)), which indicated that there were no signs of inflammation and tissue edema. The treatment of corneas with optimized formulations was exemplified in Figure 9(a), and the histopathology confirmed that no ocular irritating effects were induced by MMT-CS/TPP nanoparticle formulation compared with control group. All tested samples showed a good corneal biocompatibility which demonstrated a preferable ocular tolerance.

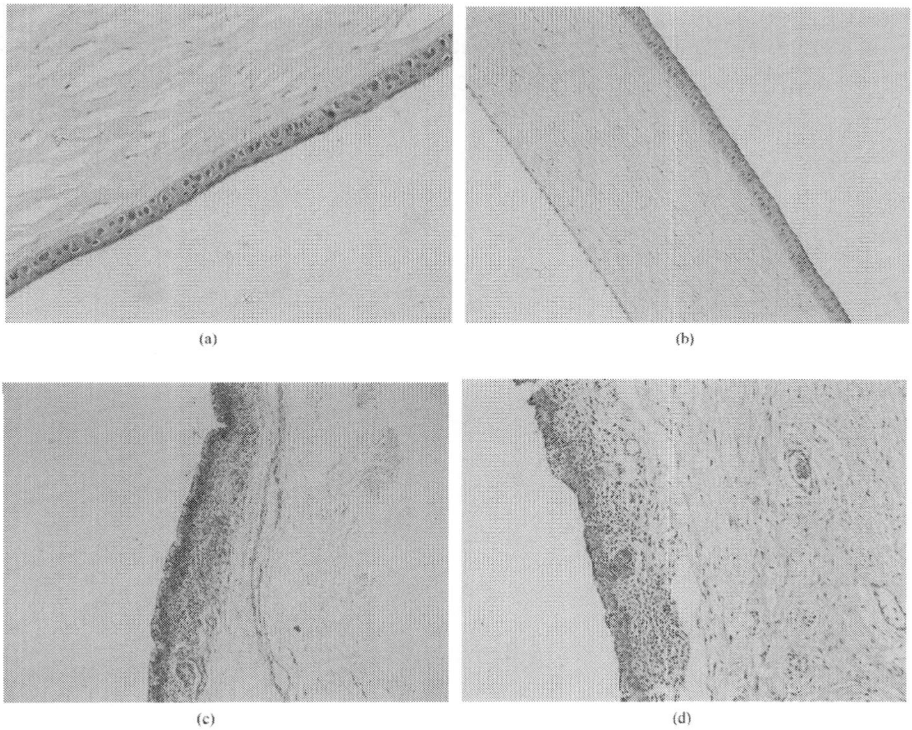

Figure 9: Histopathology microscopy of the ocular tissues including cornea (a) and (b) and conjunctiva (c) and (d) after treated with different solutions for 7 days. (a) and (c) MMT-BH/CS NPs; (b) and (d) Normal saline.

CONCLUSION

In the present study, the BH loading CS-MMT/TPP nanoparticles were prepared successfully by ionic reaction between CS and TPP based on the positively charged amino groups in CS and the negatively charged phosphate groups in TPP. In addition, the hydroxide radical on the surface of montmorillonite interacting with chitosan through hydrogen bond also played an important role during the preparation of nanoparticles. The drug-loaded nanoparticles consisting of MMT, CS and TPP, showed the enhancement of precorneal retention, sustained release, and high drug-loading capacity. The ratio between TPP and CS was critical for the control of particle size and encapsulation efficiency of nanoparticles. No ocular damage in cornea, conjunctiva or iris was observed. Consequently, these nanocarriers represented a promising approach for the circumvention of the present limitation in ocular drug delivery.

ACKNOWLEDGEMENTS

The authors would thank to the youth Natural Science Foundation for financial support (NO.51102052), and thank the eye institute of Shandong province in China for the provision of human immortalized cornea epithelial cell line (HCEC).

REFERENCES

1. Quigley, H.A. and Broman, A.T. (2006) The Number of People with Glaucoma Worldwide in 2010 and 2020. British Journal of Ophthalmology, 90, 262-267. http://www.ncbi.nlm.nih.gov/pmc/articles/PMC1856963/ http://dx.doi.org/10.1136/bjo.2005.081224
2. Mahalaxmi, D., Samarth, M.M., Shiravadekar, H.S. and Sanghavi, N.M. (1996) Stability-Indicating HPLC Method for Betaxolol HCl and Its Pharmaceutical Dosage Forms. Drug Development and Industrial Pharmacy, 22, 1037-1039. http://www.tandfonline.com/doi/abs/10.3109/03639049609065938#abstract http://dx.doi.org/10.3109/03639049609065938
3. Kaur, I.P. and Kanwar, M. (2002) Ocular Preparations: The Formulation Approach. Drug Development and Industrial Pharmacy, 28, 473-493. http://www.tandfonline.com/doi/full/10.1081/DDC-120003445#abstract http://dx.doi.org/10.1081/DDC-120003445
4. Hamalainen, K.M., Kananen, K., Auriola, S., Kontturi, K. and Urtti, A. (1997) Characterization of Paracellular and Aqueous Penetration Routes in Cornea, Conjunctiva and Sclera. Investigative Ophthalmology and Visual Science, 38, 627-634. http://www.ncbi.nlm.nih.gov/pubmed/9071216
5. Keister, J.C., Cooper, E.R., Missel, P.J., Lang, J.C. and Hager, D.F. (1991) Limits on Optimizing Ocular Drug Delivery. Journal of Pharmaceutical Sciences, 80, 50-53. http://www.ncbi.nlm.nih.gov/pubmed/?term http://dx.doi.org/10.1002/jps.2600800113
6. Calvo, C.P., Vila-Jato, J.L. and Alonso, M.J. (1997) Novel

Hydrophilic Chitosan-Polyethylene Oxide Nanoparticles as Protein Carriers. Journal of Applied Polymer Science, 63, 125-132. http://onlinelibrary.wiley.com/doi/10.1002/%28SICI%291097-4628%2819970103%2963:1%3C125::AID-APP13%3E3.0.CO;2-4/abstract http://dx.doi.org/10.1002/(SICI)1097-4628(19970103)63:1<125::AID-APP13>3.0.CO;2-4

7. Kaur, I.P. and Smitha, R. (2002) Penetration Enhancers and Ocular Bioadhesives: Two New Avenues for Ophthalmic Drug Delivery. Drug Development and Industrial Pharmacy, 28, 353-369. http://informahealthcare.com/doi/abs/10.1081/DDC-120002997%20 http://dx.doi.org/10.1081/DDC-120002997

8. Ludwig, A. (2005) The Use of Mucoadhesive Polymers in Ocular Drug Delivery. Advances Drug Delivery Reviews, 57, 1595-1639. http://www.ncbi.nlm.nih.gov/pubmed/16198021 http://dx.doi.org/10.1016/j.addr.2005.07.005

9. Mucha, M. (1997) Rheological Characteristics of Semi-Dilute Chitosan Solutions. Macromolecular Chemistry and Physics, 198, 471-484. http://onlinelibrary.wiley.com/doi/10.1002/macp.1997.021980220/abstract http://dx.doi.org/10.1002/macp.1997.021980220

10. Wang, W. and Xu, D. (1994) Viscosity and Flow Properties of Concentrated Solutions of Chitosan with Different Degrees of Deacetylation. International Journal of Biological Macromolecule, 16, 149-152. http://www.ncbi.nlm.nih.gov/pubmed/?term http://dx.doi.org/10.1016/0141-8130(94)90042-6

11. Alonso,M.J.andSanchez,A.(2003)ThePotentialofChitosaninOcular Drug Delivery. Journal of Pharmacy and Pharmacology, 55, 1451-1463. http://onlinelibrary.wiley.com/doi/10.1211/0022357022476/abstract;jsessionid=8714D64E9E7D3894BF43FBA2B0D0912C.f01t04 http://dx.doi.org/10.1211/0022357022476

12. Bhatta, R.S., Chandasana, H., Chhonker, Y.S., Rathi, C., Kumar, D., Mitra, K. and Shukla, P.K. (2012) Mucoadhesive Nanoparticles for Prolonged Ocular Delivery of Natamycin: In Vitro and Pharmacokinetics Studies. International Journal of Pharmaceutics, 432, 105-112. http://www.sciencedirect.com/science/article/pii/S0378517312004061 http://dx.doi.org/10.1016/j.

ijpharm.2012.04.060

13. Diebold, A. and Calonge, M. (2010) Applications of Nanoparticles in Ophthalmology. Progress in Retinal and Eye Research, 29, 596-609. http://www.sciencedirect.com/science?_ob=ArticleListURL&_method=list&_ArticleListID=-782671355&_sort=r&_st=13&view=c&md5=e5bfcd63708820023e9fc1c79a37ae974&searchtype=a http://dx.doi.org/10.1016/j.preteyeres.2010.08.002
14. Motwani, S.K., Chopra, S., Talegaonkar, S., Kohli, K., Ahmad, F.J. and Khar, R.K. (2008) Chitosan-Sodium Alginate Nanoparticles as Submicroscopic Reservoirs for Ocular Delivery: Formulation, Optimization and in Vitro Characterization. European Journal of Pharmaceutics and Biopharmaceutics, 68, 513-525. http://linkinghub.elsevier.com/retrieve/pii/S0939-6411(07)00308-6
15. Angela, M., Campos, D., Sanchez, A. and Alonso, M.J. (2001) Chitosan Nanoparticles: A New Vehicle for the Improvement of the Delivery of Drugs to the Ocular Surface. Application to Cyclosporin A. International Journal of Pharmaceutics, 224, 159-168. http://linkinghub.elsevier.com/retrieve/pii/S0378517301007608 http://dx.doi.org/10.1016/S0378-5173(01)00760-8
16. Ennajih, H., Bouhfid, R., Essassi, E.L.M., Bousmina, M. and El Kadib, A. (2012) Chitosan-Montmorillonite Bio-Based Aerogel Hybrid Microspheres. Microporous and Mesoporous Materials, 152, 208-213. http://dx.doi.org/10.1002/chem.201104006
17. Lin, F.H., Lee, Y.H., Jian, C.H., Wong, J.M., Schiech, M.J. and Wang, C.Y. (2002) A Study of Purified Montmorillonite Intercalated with 5-Fluorouracil as Drug Carrier. Biomaterials, 23, 1981-198. http://www.ncbi.nlm.nih.gov/pubmed/11996039 http://dx.doi.org/10.1016/s0142-9612(01)00325-8
18. Zheng, J.P., Luan, L., Wang, H.Y., Xi, L.F. and Yao, K.D. (2007) Study on Ibuprofen/Montmorillonite Intercalation Composites as Drug Release System. Applied Clay Science, 36, 297-301. http://www.sciencedirect.com/science/article/pii/S0169131707000026 http://dx.doi.org/10.1016/j.clay.2007.01.012
19. Bhatta, R.S., Chandasana, H., Chhonker, Y.S., Rathi, C., Kumar, D., Mitra, K. and Shukla, P.K. (2012) Mucoadhesive

Nanoparticles for Prolonged Ocular Delivery of Natamycin: In Vitro and Pharmacokinetics Studies. International Journal of Pharmaceutics, 432, 105-112. http://www.sciencedirect.com/science/article/pii/S0378517312004061 http://dx.doi.org/10.1016/j.ijpharm.2012.04.060

20. Qi, L.F., Xu, Z.R., Jiang, X., Hu, C.H. and Zou, X.F. (2004) Preparation and Antibacterial Activity of Chitosan Nanoparticles. Carbohydrate Research, 339, 2693-2700. http://www.ncbi.nlm.nih.gov/pubmed/15519328 http://dx.doi.org/10.1016/j.carres.2004.09.007
21. Mahmoud, A.A., El-Feky, G.S., Kamel, R. and Awad, G.E.A. (2011) Chitosan/Sulfobutylether-β-Cyclodextrin Nanoparticles as a Potential Approach for Ocular Drug Delivery. International Journal of Pharmaceutics, 413, 229-236. http://linkinghub.elsevier.com/retrieve/pii/S0378-5173(11)00358-9 http://dx.doi.org/10.1016/j.ijpharm.2011.04.031
22. Wilhelmus, K.R. (2001) The Draize Eye Test. Survey of Ophthalmology, 45, 493-515. http://www.ncbi.nlm.nih.gov/pubmed/11425356 http://dx.doi.org/10.1016/s0039-6257(01)00211-9
23. Vandenberg, G.W., Drolet, C., Scott, S.L. and Noue, J.D. (2001) Factors Affecting Protein Release from Alginate- Chitosan Coacervate Microcapsules during Production and Gastric/Intestinal Simulation. Journal of Controlled Release, 77, 297-307. http://www.sciencedirect.com/science/article/pii/S016836590100517X http://dx.doi.org/10.1016/S0168-3659(01)00517-X
24. Quan, G. and Tao, W. (2007) Chitosan Nanoparticle as Protein Delivery Carrier—Systematic Examination of Fabrication Conditions for Efficient Loading and Release. Colloid and Surfaces B: Biointerfaces, 59, 24-34. http://www.sciencedirect.com/science/article/pii/S0927776507001671
25. Kawashima, Y., Yamamoto, H., Takeuchi, H. and Kuno, Y. (2000) Mucoadhesive DL-Lactide/Glycolide Copolymer Nanoparticles Coated with Chitosan to Improve Oral Delivery of Elcatonin. Pharmaceutical Development and Technology, 5, 77-85. http://informahealthcare.com/doi/abs/10.1081/PDT-

100100522%20 http://dx.doi.org/10.1081/PDT-100100522

26. Oyarzun-Ampuero, F.A., Brea, J., Loza, M.I., Torres, D. and Alonso, M.J. (2009) Chitosan-Hyaluronic Acid Nanoparticles Loaded with Heparin for the Treatment of Asthma. International Journal of Pharmaceutics, 2, 122-129. http://www.sciencedirect.com/science/article/pii/S0378517309002129 http://dx.doi.org/10.1016/j.ijpharm.2009.04.009
27. Schiffelers, R.M., Woodle, M.C. and Scaria, P. (2004) Pharmaceutical Prospects for RNA Interference. Pharmacy Research, 21, 1-7. http://link.springer.com/article/10.1023%2FB%3APHAM.0000012145.49054.6c

CITATIONS

CHAPTER 2

Arezou Ghadi, Soleiman Mahjoub, Fatemeh Tabandeh, Farid Talebnia 2014 Synthesis and optimization of chitosan nanoparticles: Potential applications in nanomedicine and biomedical engineering, Caspian J Intern Med 2014; 5(3): 156-161.

CHAPTER 3

Vivek Kamat, Dhananjay Bodas & Kishore Paknikar, 2016, Chitosan nanoparticles synthesis caught in action using microdroplet reactions, Scientific Reports | 6:22260 | DOI: 10.1038/srep22260

CHAPTER 4

Reynaldo Esquivel, Josué Juárez, Mario Almada, Jaime Ibarra, andMiguel A. Valdez 2015 Synthesis and characterization of new thiolated chitosan nanoparticles obtained by ionic gelation method, Hindawi Publishing Corporation International Journal of Polymer ScienceVolume 2015, Article ID 502058, 18 pages http://dx.doi.org/10.1155/2015/502058

CHAPTER 5

Tan Vui Nguyen, Thi Thu Ha Nguyen, San-Lang Wang, Thi Phuong Khanh Vo, Anh Dzung Nguyen, 2015 Preparation of chitosan nanoparticles by tpp ionic gelation combined with spray drying, and the antibacterial activity of chitosan nanoparticles and a chitosan nanoparticle–amoxicillin complex, Res Chem Intermed DOI 10.1007/s11164-016-2428-8

CHAPTER 6

Sijumon Kunjachan, Sajan Jose, Twan Lammers, 2015 Understanding the mechanism of ionic gelation for synthesis of chitosan nanoparticles using qualitative techniques,

Asian Journal of Pharmaceutics - April-June 2010 DOI: 10.4103/0973-8398.68467 http://www.asiapharmaceutics.info/index.php/ajp/article/view/220

CHAPTER 7

M.Abd Elgadir, Md.Salim Uddin, Sahena Ferdous, Aishah Adam, Ahmed Jalal Khan Chowdhury, Md.Zaidul Islam Sarker, 2015, Impact of chitosan composites and chitosan nanoparticle composites on various drug delivery systems, Journal of Food and Drug Analysis, 23 (4), pp. 619-629.

CHAPTER 8

Facchi DP, Facchi PF, Martins AF (2016) N,N,N-Trimethyl Chitosan and Its Potential Bactericidal Activity: Current Aspects and Technological Applications. J Infect Dis Ther 4: 291. doi:10.4172/2332-0877.1000291

CHAPTER 9

Patel R, Gajra B, Parikh RH, Patel G (2016) Ganciclovir Loaded Chitosan Nanoparticles: Preparation and Characterization. J Nanomed Nanotechnol 7: 411. doi: 10.4172/2157-7439.1000411

CHAPTER 10

Manikandan A, Sathiyabama M (2015) Green Synthesis of Copper-Chitosan Nanoparticles and Study of its Antibacterial Activity. J Nanomed Nanotechnol 6: 251. doi: 10.4172/2157-7439.1000251

CHAPTER 11

Cintia Rodrigues Maruyama, Mariana Guilger, Mônica Pascoli, Natalia Bileshy- José, P.C. Abhilash, Leonardo Fernandes Fraceto & Renata de Lima1,3 Scientific Reports | 6:19768 | DOI: 10.1038/srep19768

CHAPTER 12

Samzadeh-Kermani, A. and Esfandiary, N. (2016) Synthesis and Characterization of New Biodegradable Chitosan/Polyvinyl Alcohol/Cellulose Nanocomposite. Advances in Nanoparticles, 5, 18-26. doi: 10.4236/anp.2016.51003.

CHAPTER 13

Parida, U. , Rout, N. and Bindhani, B. (2013) In vitro properties of chitosan nanoparticles induce apoptosis in human lymphoma SUDHL-4 cell line. Advances in Bioscience and Biotechnology, 4, 1118-1127. doi: 10.4236/abb.2013.412148.

CHAPTER 14

Hou, D. , Gui, R. , Hu, S. , Huang, Y. , Feng, Z. and Ping, Q. (2015) Preparation and Characterization of Novel Drug-Inserted-Montmorillonite Chitosan Carriers for Ocular Drug Delivery. Advances in Nanoparticles, 4, 70-84. doi: 10.4236/anp.2015.43009.

INDEX